CfE Higher MATHS

GRADE C BOOSTER

CfE Higher MATHS
GRADE C BOOSTER

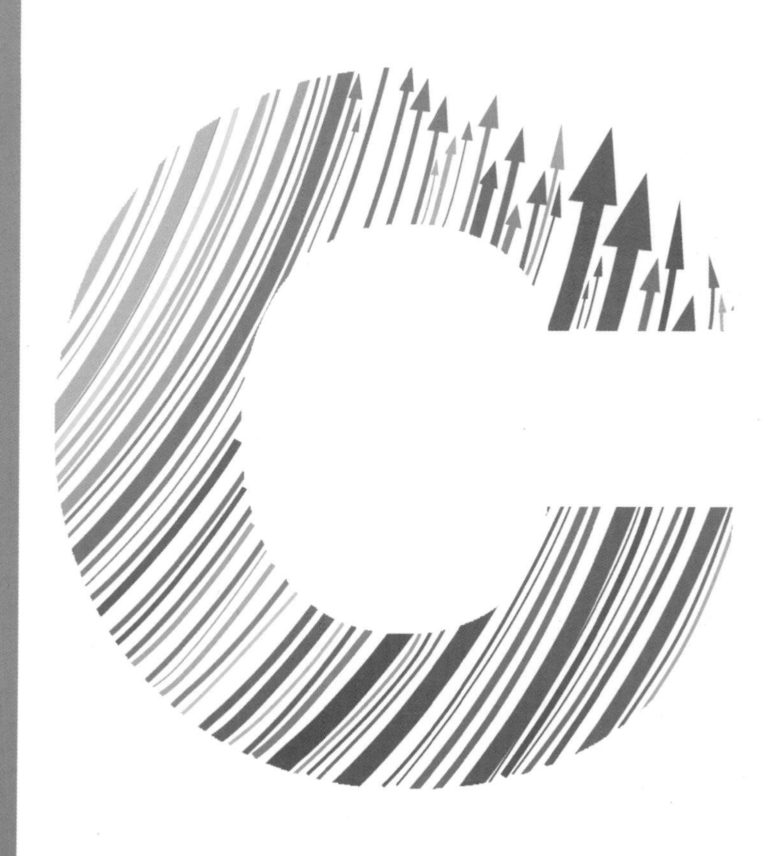

Stuart Welsh

001/06112015

10 9 8 7 6 5 4 3 2 1

ISBN 9780007590827

Published by
Leckie & Leckie Ltd
An imprint of HarperCollins*Publishers*
Westerhill Road, Bishopbriggs, Glasgow, G64 2QT
T: 0844 576 8126   F: 0844 576 8131
leckieandleckie@harpercollins.co.uk
www.leckieandleckie.co.uk

Publisher: Peter Dennis
Project manager: Craig Balfour and Keren McGill

Special thanks to
Jill Laidlaw (copy edit)
Louise Robb (proofread)
Caleb O'Loan (proofread)
Jouve India (layout and illustration)
Ink Tank (cover)

Printed in China

A CIP Catalogue record for this book is available from the British Library.

Acknowledgements
All images © Shutterstock.com

Contents

Introduction to CfE Higher Maths

How to use this book

Who is this book for?

This book has been designed to provide support for candidates who are looking to give themselves the best possible chance of securing a pass in CfE Higher Mathematics. It focuses specifically on helping you understand and be able to manipulate the basic skills and knowledge of the Higher Maths course to give you the best chance of achieving an overall course award. However, this book covers all aspects of the Higher Maths course and is useful for all candidates.

What's in this book?

This book contains original questions along with questions from SQA specimen, exemplar and past examination papers. It includes a description of how to answer the questions, a guide to show you where marks (✓) are awarded (all examples in Chapter 6 and the first two examples in Chapter 9 do not have marks assigned as they are not strictly at Higher level, though do contain essential skills), and hints to make sure you pick up as many marks as possible. The book follows the SQA course structure and is divided into three units with four chapters in each. Within each chapter you will find an introduction to the topic, a list of essential prior knowledge and a mixture of short and extended response questions. In some chapters you will find integrative questions, which pull together knowledge and skills from other parts of the course. This book also contains a useful glossary of mathematical terms that are frequently used in exams, to help you understand what questions are actually asking. Glossary words will be blue and in bold the first time they appear in each chapter.

Get the basics right

Higher Mathematics is a challenging course. It draws upon mathematical skills and knowledge developed over years of studying mathematics. National 5 provides good preparation for the Higher course and previous knowledge of algebra, geometry, arithmetic and trigonometry is strongly desirable. Chapter 6 covers essential mathematical skills. You should work at mastering the skills in Chapter 6 before tackling the rest of the book because, while not strictly being at Higher level, if you do not understand these skills then valuable marks can be lost creating a barrier to demonstrating the Higher material you have worked hard to become confident with.

When to use this book

This book is suitable for you to use from the very start of your Higher Mathematics course and will give you a valuable opportunity to see examination style questions at the same time as you learn a topic in class. As you work through the examples you might meet questions that seem unfamiliar. This could be because you have not yet covered either this topic or this specific question type in class. Check with your teacher to see if this is the case. It could be that the wording or context of the question is unfamiliar. Look out for keywords and use the glossary to help. You will find that in most cases, despite looking tricky at first, the question will be solved using techniques you are familiar with. You can also use this book to help you prepare for the final examination and you should use it along with your class notes, textbooks, practice paper books and any other resources you have as part of your preparation for the final examination.

The CfE Higher Maths course

The CfE Higher Mathematics course is relatively new, although the content is largely the same as the old Higher course. An overview of the course structure, the structure of the final examination, and the types of questions you will be expected to answer is given in Chapters 2–4. All the information included in this book has been based on SQA guidance and was correct at the time of going to press. However, changes may be made and up-to-date details of the course, along with other useful information can be downloaded from the SQA CfE Higher Mathematics webpage. The easiest way to find this is to do an internet search for 'SQA CfE Higher Maths'.

Alternative methods

Many of the questions in Higher Mathematics can be tackled in different ways and everybody will have their own preferred method. The methods in this book were chosen because they are the most accessible/easiest to understand and because many are transferrable, i.e. the same/or a similar method can be used in questions from more than one topic. Always use a method you are familiar with and remember that your teacher may have shown you a particular method because they felt it was the most appropriate one for you.

The structure of the CfE Higher Mathematics course

The essentials

The CfE Higher Mathematics course has been divided by the SQA into three units. These units are called: Expressions and Functions; Relationships and Calculus; and Applications. Each unit contains four separate topics. The course does not have to be taught in any particular order, although it makes sense to complete certain topics before others.

Full details of the course and other useful information can be downloaded from the SQA CfE Higher Mathematics webpage. You can find this by searching the internet for 'CfE Higher Maths' or by keying http://www.sqa.org.uk/sqa/47910.html into your web browser.

The CfE Higher Maths course builds on the skills, knowledge and understanding required for National 5 Mathematics. Following successful completion of the Higher Maths course you could go on to study Advanced Higher Maths, Advanced Higher Statistics, Advanced Higher Mechanics or you could go on to Further or Higher Education to study a wide variety of courses involving maths.

In the CfE Higher Maths course there is an emphasis on skills development and application of skills in a problem solving context. To gain an overall course award you will need to pass all three of the (internally assessed) units mentioned earlier. To pass Higher Maths you need to demonstrate you are confident with the basic processing skills required for each unit, plus you will have to show that you have developed mathematical reasoning and communication skills. This is done by interpreting a situation where maths can be used, identifying a suitable strategy and explaining a solution, often by relating it to context. In addition to passing the units, you will also have to pass the final examination.

There is no assignment for Higher Mathematics.

The structure of the CfE Higher Mathematics examination

The final examination

Your final examination will be set and marked by the SQA. When the examiners are writing the paper, they try to include content from across the whole course. The examination is not long enough to ask you every possible type of question, although you should be prepared to answer questions on any topic. See Chapter 4 (Types of questions) for more information on the style and structure of the question you will be expected to answer.

Your final examination consists of two papers. You will sit both of these papers on the same day, with a short interval in between.

Paper 1

Paper 1 is 70 minutes long and is to be completed without the use of a calculator.

The paper is worth 60 marks and will consist of short and extended response questions.

You are required to show an understanding of underlying processes, which means you can expect non-calculator questions involving solving equations or working with surds, indices and **logarithms**.

Paper 2

Paper 2 is 90 minutes long and a calculator may be used.

The paper is worth 70 marks and will consist of short and extended response questions.

You won't need your calculator for all the questions in paper 2. In fact, using a calculator to work out values that could be left as **exact values** can lead to rounding errors. However, a calculator allows more opportunity for application and reasoning, and is helpful in situations where complex calculations are required to solve problems.

Proportion of level 'C' questions

The SQA has stated that approximately 65 % of the marks will be available for 'C' level responses. Some questions will use a 'stepped' approach, i.e. the question will be split into parts to allow you the opportunity to demonstrate your ability at 'C' level.

Your overall grade will be based on your performance in the course assessment (final examination). To be awarded a grade 'C' you must demonstrate successful performance in all of the units, plus demonstrate *successful performance* in relation to the skills, knowledge and understanding required for the course. For a grade 'A' you must demonstrate successful performance in all of the units plus demonstrate *a consistently high level of performance* in relation to the skills, knowledge and understanding required for the course.

A specimen paper, exemplar paper, plus other useful information (including past examination papers once they have been sat), can be downloaded from the SQA CfE Higher Mathematics webpage. The easiest way to find this is to do an internet search for 'CfE Higher Maths'.

Types of questions

Short response questions

These questions usually test a skill or process. You can expect these questions to be worth between 3 and 5 marks each. Some may be presented in two parts where your answer to part (a) will probably (but not always) be required to answer part (b).

Extended response questions

These questions are longer and usually combine knowledge and skills from across the course. You can expect these questions to be worth between 6 and 10 marks in total, sometimes even more. Extended questions are likely to be divided into at least two parts. Some will be divided into as many as four parts with the most challenging part often towards the end of the question. Look for connections between the parts of a question: these are almost always linked and your answer to part (a) or part (b) may well be needed later in the question. It is also worth remembering that you don't always have to be able to do part (a) of a question to be able to go on and attempt later parts of a question. So don't give up if you can't do part (a), take a deep breath and keep going!

Strategy and communication

It is possible that a question in paper 1 or paper 2 might require you to explain a strategy or justify an answer. To get as many marks as possible for this type of question you will need to show **all stages of your working** and communicate clearly using appropriate mathematical vocabulary and notation. When using standard results it is not good enough to simply quote the result, for example, when using **gradients** to show lines are perpendicular, it is not enough to state: $m_1 \times m_2 = -1$. Instead you should involve the values you have calculated in the question, e.g.

$$m_{AB} \times m_{CD} = \frac{3}{4} \times -\frac{4}{3} = -1, \text{ hence AB is perpendicular to CD}$$

Use the glossary of mathematical terms and symbols to help you become more familiar with the language of Higher Mathematics.

Formulae

A number of mathematical formulae are provided for you in your examination. They are usually printed inside the front cover of your question paper. You should make sure you are confident using all of these formulae. When you use a formula it is important that you link it to the context or diagram given in the question, for example, in a **vectors** question involving vectors **p** and **q**, using the **scalar product** formula with **a** and **b** will not guarantee all the marks. You must replace the letters in the formula with the letters you are given in the question. The formulae given in your examination are included in Chapter 7 of this book, along with some other useful formulae not provided in your examination, which you should memorise.

Definitions of mathematics-specific command words

You will encounter various command words in your examination which have specific mathematical meanings. The list below was produced by the SQA to give guidance as to how these words should be interpreted.

- Determine: obtain an answer from given facts, figures or information.
- Expand: multiply out an algebraic **expression** by making use of the distributive law or a compound trigonometric expression by making use of one of the addition formulae for $\sin(A \pm B)$ or $\cos(A \pm B)$.
- Express: use given information to rewrite an expression in a given form.
- Find: obtain an answer showing relevant stages of working.
- Hence: use the previous answer to proceed.
- Hence, or otherwise: use the previous answer to proceed. However, another method may also be used.
- Identify: provide an answer from a number of possibilities.
- Justify: give good reasons for the conclusion(s) reached.
- Show that: use mathematics to prove something, e.g. that a statement or given value is correct – all steps, including the required conclusion, must be shown.
- Sketch: give a general idea of the required shape or relationship and annotate with all relevant points or features.
- Solve: obtain the answer(s) using algebraic and/or numerical and/or graphical methods.

General advice

Comments and hints

Throughout this book you will find comments and hints designed to help you get as many marks as possible in your Higher Mathematics examination. Some hints are specific to certain topics and question types. More general advice is included in this chapter.

Diagrams

Some questions in your final examination will contain a diagram. It might be helpful if you re-draw the diagram on your answer booklet. This will let you mark on any additional information from the question, and any results you calculate. If you don't want to spend time re-drawing, just write directly on the diagram on the question paper but remember that anything you write on the question paper will not be seen by an examiner and won't get you any marks. Some questions might not have a diagram. It is often very helpful to make a sketch of the situation; this makes it much easier to see what is going on, especially in geometry-based questions, i.e. straight lines, circles and **vectors**.

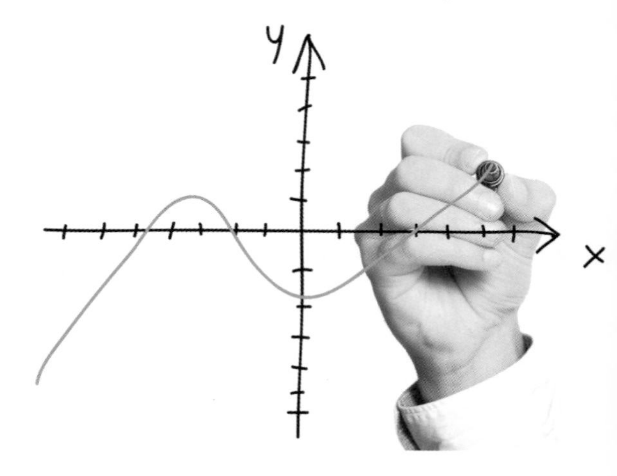

Don't get tied in knots

When you are working through a question you might find that the numbers or algebra quickly become very complicated. In these cases you should consider whether you might have made a simple error earlier on; perhaps you have copied down the question incorrectly, or made a simple arithmetic error. Instead of slogging it out with some nasty fractions, go back to the start, make sure you have used the correct numbers and carefully check through your working for errors.

Play to your strengths

In your examination you will find some questions harder than others. Try not to get stuck on one question for too long as you may run out of time. If you feel that you have come to a dead-end with a question, it is better to move on and return to that question later. If you find you are stuck with a question early in the paper, this doesn't mean that all the questions that come after will be even harder; in fact, you might find that some of the later questions seem easier to you.

Use your time wisely

If you have spare time at the end of the examination, use it to check over your answers. Ask yourself whether you have fully answered the question, e.g. if the question asked for coordinates, have you given coordinates or just the value of x and/or y? If you have sketched a graph, have you marked on **all** the relevant points? If you change your mind about an answer, don't score it out until you have replaced it with something you are confident with. Remember, wrong answers can be awarded 'follow-through' marks and can sometimes get almost all the marks for a question provided you have shown all working and have communicated your thinking clearly.

Always fully simplify

In the Higher Mathematics examination, you are expected to leave all your answers in a fully simplified form. It would be a shame to throw away precious marks because you didn't simplify a fraction. Values like $\sqrt{25}$ must be simplified to 5. You should also get into the habit of simplifying surds, for example, $\sqrt{12} = 2\sqrt{3}$, and you must make sure that values like 6^2 are left as 36. **Logarithms** should also be simplified as far as possible, for example, $\log_2 8 = 3$.

Keep it neat

Take the time to keep your working as neat as you can. You will often have to look back at previous working as you progress through a question so don't throw away marks by misreading your own writing, i.e. 7's can look like 1's, 8's can become 6's, 4's can turn into 9's etc. You also risk losing marks for working that is ambiguous or is impossible to read. Use a ruler for any sketches or graphs. You might find that keeping sketches and graphs reasonably small makes them neater and easier to draw.

Know your calculator

Make sure you use the same calculator throughout the course and in the final examination. It is important that you are familiar with the useful features of your calculator and can use it to your best advantage. Instruction manuals can be found on the internet for most models and you might be surprised to find out what your calculator can actually do! Graphic calculators can be useful in checking that you have drawn graphs correctly, although they can be expensive and complicated to work. Try to use a calculator which displays more than one line and allows for true mathematical input/output, i.e. fractions, indices etc. are displayed in the same way as you would write them down, for example,

$$\cos^{-1}\left(\frac{14}{\sqrt{35}\sqrt{45}}\right)$$

69.34335882

This will help make sure you enter calculations correctly. Watch out for questions that require you to give answers in **radians**. You will need to switch the mode on your calculator from degrees to radians, make sure you are comfortable doing this and don't forget to *switch it back!*

This book contains a mixture of questions from past calculator and non-calculator papers. It is important to remember that questions in the calculator papers don't necessarily need a calculator to solve them. Many of them are what the SQA call 'calculator neutral' and are often algebra- or geometry-based.

In this book this icon beside a question means that a calculator is allowed. This icon means that you should be able to answer the questions without the use of a calculator.

Essential mathematical skills

Get the basics right

Higher Mathematics is a challenging course. It builds upon mathematical skills and knowledge developed over years of studying Maths. National 5 provides good preparation for the Higher course and previous knowledge of algebra, geometry, arithmetic and trigonometry is strongly desirable. This chapter covers skills which, while not always being considered Higher skills, will help you secure more marks in the final exam. You should make sure you have mastered the skills in this chapter before you begin working through the rest of the book.

Number skills

The best way to improve your number skills is to do as much non-calculator number work as you can. Fight the urge to reach for your calculator and have a go yourself. There is no need for you to do calculations 'in your head', set out some working and take as many steps as you need to get it right.

Powers

Good knowledge of square numbers and other powers will help you deal with surds, indices, **differentiation, integration, logarithms** and **exponentials**.

Some commonly encountered square numbers:

1^2	2^2	3^2	4^2	5^2	6^2	7^2	8^2	9^2	10^2	11^2	12^2	15^2	20^2	100^2
1	4	9	16	25	36	49	64	81	100	121	144	225	400	10000

You should be familiar with the first few powers of 2:

2^0	2^1	2^2	2^3	2^4	2^5	2^6
1	2	4	8	16	32	64

It can also be helpful to know a few other frequently occurring powers:

3^3	3^4	4^3	5^3	10^3
27	81	64	125	1000

Integers (positive and negative numbers)

Careless errors with negatives can cause you to lose several marks. Remember the rules for multiplying and dividing integers:

positive × positive = positive

positive × negative = negative

negative × positive = negative

negative × negative = positive

positive ÷ positive = positive

positive ÷ negative = negative

negative ÷ positive = negative

negative ÷ negative = positive

Squaring a negative number will give a positive answer, e.g. $(-4)^2 = 16$.

Cubing a negative number will give a negative answer, e.g. $(-3)^3 = -27$.

In general:

A negative number raised to an *odd* power will always give a *negative* answer, e.g. $(-2)^5 = -32$.

A negative number raised to an *even* power will always give a *positive* answer, e.g. $(-2)^6 = 64$.

There are no set rules when you are adding and subtracting integers; just think about the number you are starting with and decide whether adding or subtracting a positive or negative number will give you an answer that is above or below zero. Take your time and always check that your answer makes sense.

Example 6.1

Q Evaluate $-12 + (-5)$.

A

$$-12 + (-5)$$
$$= -12 - 5$$
$$= -17$$

Starting with a negative number and adding another negative number means the answer must be an even larger negative number than the one you started with.

Example 6.2

Q Evaluate $-7 - (-4)$.

A

$$-7 - (-4)$$
$$= -7 + 4 = -3$$

The 'double negative' acts like a plus but be careful, in this case we are not adding enough to give a positive answer.

Substituting negatives

You must always use brackets when substituting negatives into an **expression** otherwise you risk losing marks.

Example 6.3

Find $f(-3)$ if $f(x) = x^2 - x$.

$f(-3) = (-3)^2 - (-3)$

$f(-3) = 9 + 3$

$f(-3) = 12$

You should always use a line of working just to substitute in the values, don't try to work anything out at this stage. Writing -3^2 is not just poor communication, it is actually wrong as $-3^2 = -9$. The brackets must be there to make sure the '–' is also squared.

Remember, squaring a negative always gives a positive. Also notice that the double negative has become a plus.

Fractions

As part of your response to larger questions you will be expected to be able to find a fraction of an amount, as well as add, subtract, multiply and divide fractions without a calculator. Mistakes here can quickly make your answers to trigonometry and calculus questions very messy.

Example 6.4

Find $\frac{1}{4} \div \frac{3}{2}$.

$\frac{1}{4} \div \frac{3}{2}$

$= \frac{1}{4} \times \frac{2}{3}$

$= \frac{1}{\cancel{4}^2} \times \frac{\cancel{2}^1}{3}$

$= \frac{1}{2} \times \frac{1}{3}$

$= \frac{1}{6}$

To divide by a fraction, swap the numerator and denominator and multiply instead.

Always look to simplify first by cancelling diagonally. This will keep the numbers small and save you from having to simplify later on.

Multiply numerators and multiply denominators; there is no need for a common denominator.

Example 6.5

Q Evaluate $3\frac{2}{9}+1\frac{5}{6}$.

A

$$3\tfrac{2}{9}+1\tfrac{5}{6}$$
$$=4\tfrac{2}{9}+\tfrac{5}{6}$$
$$=4\tfrac{4}{18}+\tfrac{15}{18}$$
$$=4\tfrac{19}{18}$$
$$=5\tfrac{1}{18}$$

Add or subtract any whole numbers first.

Find the *lowest* common denominator. If you use a different common denominator you will have to simplify your answer later.

Add the numerators but keep the denominator the same.

Convert any top heavy fractions to mixed numbers.

Example 6.6

Q Evaluate $-16\left(-\frac{3}{4}\right)$.

A $-16\left(-\frac{3}{4}\right)=12$

Stay switched on! Numbers beside each other in brackets should be multiplied. Multiplying two negatives here will give a positive answer. Find $\frac{3}{4}$ of 16 by dividing by 4 then multiplying by 3.

Rounding

You should be able to round to a given number of significant figures or decimal places. This is important in questions involving logarithms, exponentials, **sequences** and calculus where a correctly rounded answer is sometimes needed in a later part of the question.

Example 6.7

Q Round 69354000 to three significant figures.

A $69354000 \approx 69400000$ to three significant figures

In this example the number has been rounded to the nearest 'hundred thousand'. The ≈ symbol means 'approximately equal to'.

Example 6.8

Q Round 0.0039472 to three significant figures.

A $0.0039472 \approx 0.00395$ to three significant figures

Leading zeros are not significant. Notice that rounding 0.0039472 to three *decimal places* would give 0.004.

Algebra skills

Being confident when working with algebraic expressions and equations will go a long way toward securing a good pass in Higher Mathematics. Your algebraic skills are tested throughout the course and the SQA has previously stated that approximately 30–45 % of the marks in your final examination will be available for correctly performing algebraic routines. As with most mathematics, if you learn the rules and apply them in the correct order you will have no problem.

Factorising

Always check your answer when **factorising** by multiplying back out. If you do this you should never leave a wrong answer. Look out for (in this order):

A common factor

Example 6.9

Factorise $3x^3 + 12x^2$.

$3x^3 + 12x^2$
$= 3x^2(x + 4)$

Look for numbers and letters that are **factors** of every term. Always make sure you have removed the highest common factor. You can check this by examining what is in the brackets – if it doesn't factorise further, you are finished.

A difference of two squares

Example 6.10

Factorise $7 - 28x^2$.

$7 - 28x^2$
$= 7(1 - 4x^2)$
$= 7(1 - 2x)(1 + 2x)$

Always remove any common factors first. Here the bracket contains a difference of two squares which will factorise further. Don't forget that 1 is a square number.

Check multiplying this back out gives the original expression.

A trinomial (quadratic)

Example 6.11

Factorise $x^2 - x - 12$.

When factorising a trinomial where the term in the middle is $+x$ or $-x$, the numbers you choose for the end of the brackets will always differ by one.

$x^2 - x - 12$
$= (x + 3)(x - 4)$

Always check that multiplying back out gives the original expression.

Example 6.12

Q Factorise $6 - x - 2x^2$.

A
$$6 - x - 2x^2$$
$$= (3 - 2x)(2 + x)$$

If the trinomial has a negative x^2 term, write it with the number term at the start and the x^2 at the end, as above, then factorise as normal and check, check, check!

Solving equations

Questions that require you to find: where a graph crosses the x or y axis; the **value** of a **function** at a given point; the points of **intersection** of two lines or curves; a missing value from a coordinate; the coordinates of a **stationary point**; or any missing values of x, y, k, p etc., will almost always require you to solve some form of equation. Factorising is often required before equations can be solved. In other cases, careful manipulation of fractions, squares and roots will be needed.

Hint

Setting equations equal to zero isn't always the best approach. Equations which don't require factorising need to be solved slightly differently.

Example 6.13

Q Solve $16x + 8 = 0$.

A
$$16x = -8$$
$$x = -\frac{1}{2}$$

Subtract 8 from both sides but don't take your eye off the ball at this stage! It is easy to see 16 and −8 and think the answer is −2. This is a reflex thought based on number bonds you formed years ago.

Of course, here we are dividing −8 by 16 giving an answer of $-\frac{1}{2}$. Make sure you concentrate right to the very end when solving equations.

Example 6.14

Solve $6x^2 - 26x - 20 = 0$.

$6x^2 - 26x - 20 = 0$

$2(3x^2 - 13x - 10) = 0$

$2(3x + 2)(x - 5) = 0$

$3x + 2 = 0$ or $x - 5 = 0$

$3x = -2$ or $x = 5$

$x = -\frac{2}{3}$ or $x = 5$

Always look first for a common factor. In this example it is possible to divide every term by 2. Doing this won't affect the solution(s) but will make factorising easier.

Check your factorising is correct by multiplying back out and don't forget the '= 0'.

Set *every* factor equal to zero.

Solve for x. Take as many steps as you need and be careful with negatives and fractions.

Hint

When solving an equation like the one above you are finding the value(s) of x at which the graph of $y = 6x^2 - 26x - 20$ crosses the x-axis. These points are known as **'roots'**, 'solutions' or 'zeros' and are shown on this diagram.

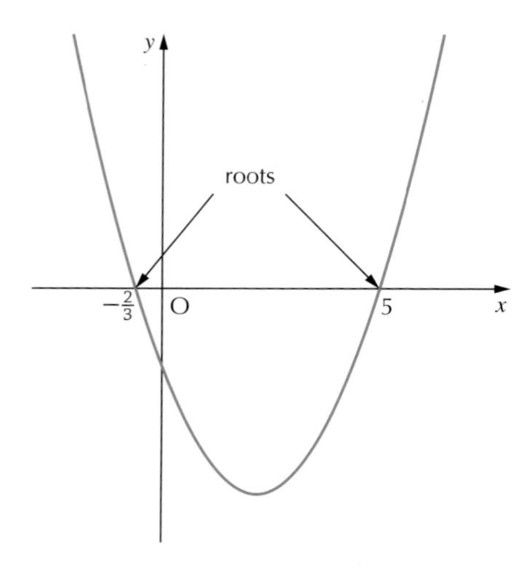

If the graph does cross the x – axis (not all do), this will happen when $y = 0$. Don't miss out the '= 0' or you may lose a mark.

Example 6.15

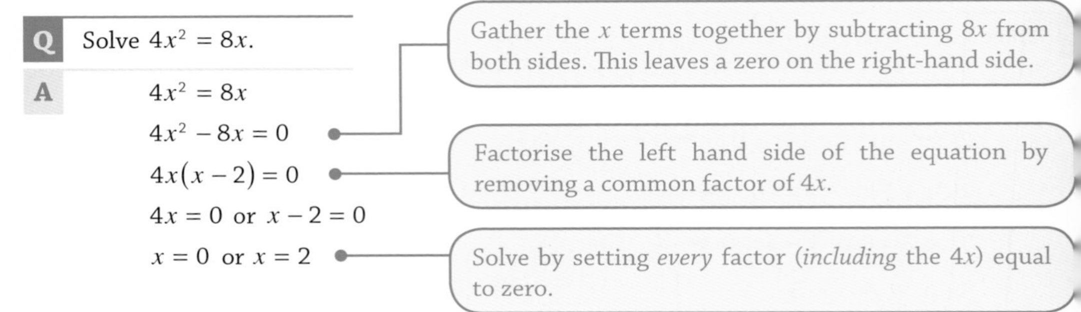

Solving inequalities

Inequalities (sometimes called inequations) often appear when examining the nature of roots using the **discriminant** or when considering the **domain** of a function. See Chapter 10 (Functions and graphs) for more on the domain of a function and Chapter 12 (Polynomials) for more on the nature of roots. To solve inequalities correctly you must make sure you are confident when you should (and shouldn't) reverse the inequality sign.

Example 6.16

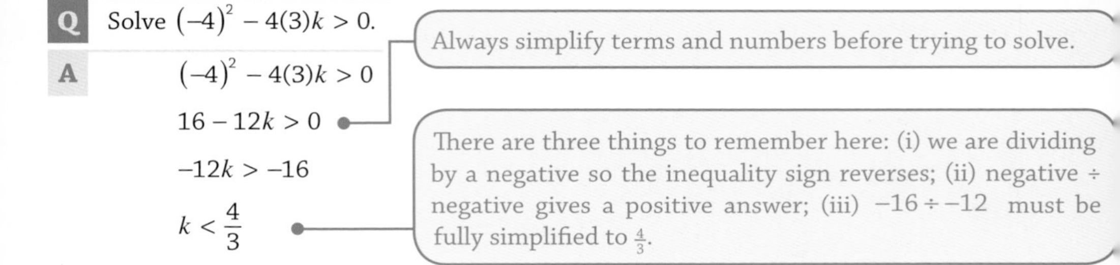

Simultaneous equations

One use of simultaneous equations is to find where two straight lines intersect (cross). However, simultaneous equations can also be helpful in questions involving polynomials or sequences where you are required to find more than one missing value. You might be familiar with the elimination method, although a substitution method can often be more straightforward, especially if you can easily change the subject of one of the equations to x or y.

Example 6.17

Find the coordinates of the point of intersection of the lines $3y = x + 11$ and $y = -3x + 7$.

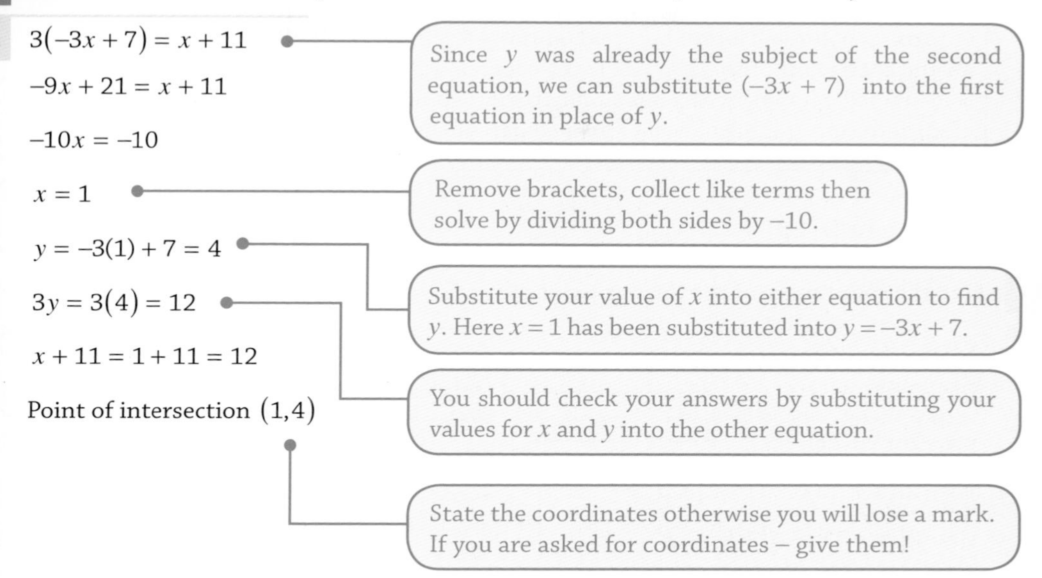

Completing the square

This technique is part of the National 5 course and is useful when you are trying to find the coordinates of the maximum or minimum **turning point** of a **parabola**. Completing the square can also be used to solve quadratic equations instead of using the quadratic formula. At Higher level you will be expected to deal with equations where the **coefficient** of x^2 is a number other than 1.

Example 6.18

Find the coordinates and nature of the turning point of $y = x^2 - 6x + 11$.

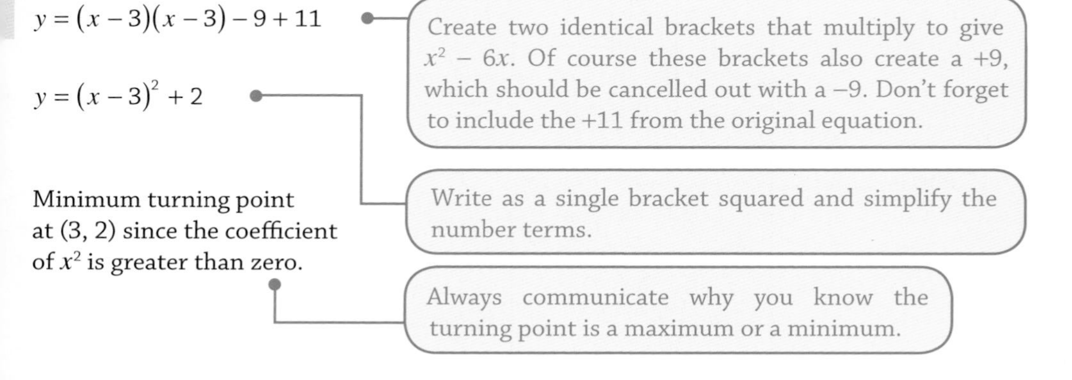

Hint

A turning point will either be a maximum or a minimum. The 'nature' is controlled by the coefficient of the x^2 term. If the coefficient is less than zero, the graph will be a ∩-shaped parabola with a maximum turning point. If the coefficient of x^2 is greater than zero, the graph will be a ∪-shaped parabola with a minimum turning point.

Example 6.19

Q Find the coordinates of the turning point of $y = 5 - 4x - 2x^2$.

A

$y = 5 - 4x - 2x^2$

$y = -2x^2 - 4x + 5$ — Re-write the equation with the x^2 term first. Take care not to change any signs.

$y = -2(x^2 + 2x) + 5$ — Take out a common factor from the first two terms only, in this case we have divided by −2. Notice how the x term has become $+2x$ since $-4x \div -2 = 2x$.

$y = -2[(x + 1)(x + 1) - 1] + 5$ — Create two identical brackets which multiply to give $x^2 + 2x$. Of course these brackets also create a +1, which should be cancelled out with a −1 that must be inside the square brackets. Don't forget to include the +5 from the original equation outside the square brackets.

$y = -2[(x + 1)^2 - 1] + 5$ — Write as a single bracket squared.

$y = -2(x + 1)^2 + 2 + 5$ — Multiply everything *inside* the square bracket by the −2. Note that the 5 is not multiplied.

$y = -2(x + 1)^2 + 7$ — Simplify the number terms.

Maximum turning point at (−1, 7). — We know it is a maximum TP because the original x^2 term has a negative coefficient giving a ∩-shaped parabola. Note that the −2 in front of the bracket only affects the shape of the curve and has no effect on the turning point.

Surds

Surds are part of the National 5 Mathematics course and being skilled in working with surds is essential for Higher Mathematics. You should be able to simplify, multiply, divide and evaluate surds. You should also be able to rationalise the denominator of a surd.

The main rules for multiplying and dividing surds are given here:

$\sqrt{a} \times \sqrt{b}$	$= \sqrt{ab}$	When surds are being multiplied they can be written as a single surd.
$c\sqrt{a} \times d\sqrt{b}$	$= cd\sqrt{ab}$	Coefficients on surds are multiplied as normal.
$\frac{\sqrt{a}}{\sqrt{b}}$	$= \sqrt{\frac{a}{b}}$	When surds are being divided they can be written as a single surd.
$\frac{c\sqrt{a}}{d\sqrt{b}}$	$= \frac{c}{d}\sqrt{\frac{a}{b}}$	Coefficients on surds are divided as normal.

Simplifying surds

Surds can be simplified by looking for square number factors. Always try to find the highest square number factor otherwise you will need to simplify again.

Example 6.20

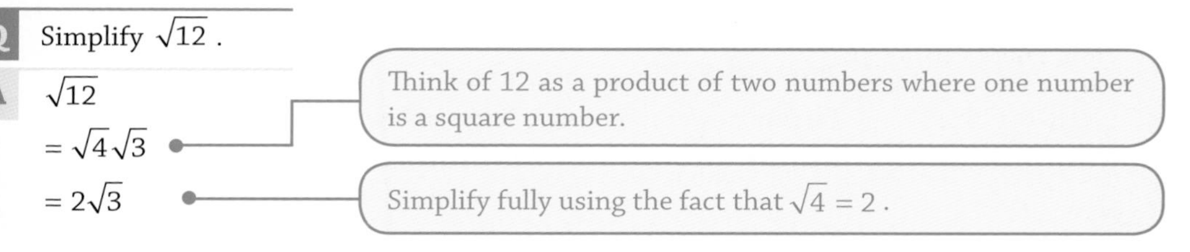

Q Simplify $\sqrt{12}$.

A $\sqrt{12}$

$= \sqrt{4}\sqrt{3}$ — Think of 12 as a product of two numbers where one number is a square number.

$= 2\sqrt{3}$ — Simplify fully using the fact that $\sqrt{4} = 2$.

Rationalising the denominator of a surd

A surd is not fully simplified if it is left with an **irrational** denominator, i.e. where the denominator contains a surd. Being able to remove irrational denominators can make solving certain equations easier.

Example 6.21

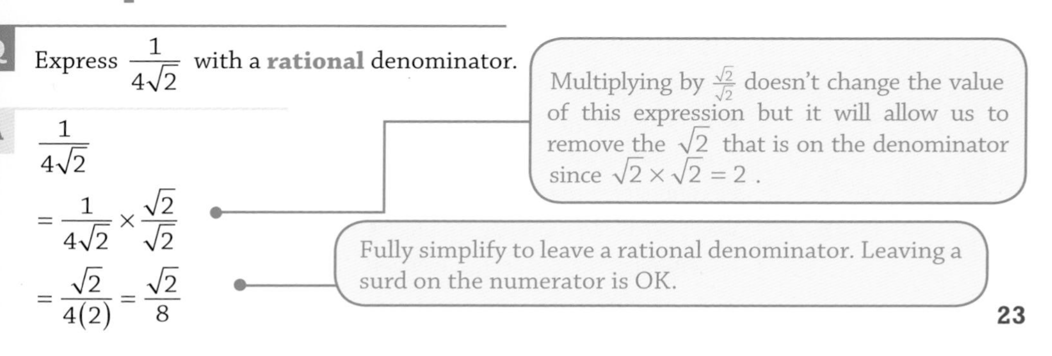

Q Express $\frac{1}{4\sqrt{2}}$ with a **rational** denominator.

A $\frac{1}{4\sqrt{2}}$

$= \frac{1}{4\sqrt{2}} \times \frac{\sqrt{2}}{\sqrt{2}}$ — Multiplying by $\frac{\sqrt{2}}{\sqrt{2}}$ doesn't change the value of this expression but it will allow us to remove the $\sqrt{2}$ that is on the denominator since $\sqrt{2} \times \sqrt{2} = 2$.

$= \frac{\sqrt{2}}{4(2)} = \frac{\sqrt{2}}{8}$ — Fully simplify to leave a rational denominator. Leaving a surd on the numerator is OK.

Simplifying expressions using the laws of indices

Skills in working with indices are essential for Higher Mathematics. Indices could appear anywhere in the course and are commonly found in calculus questions. You should be able to simplify expressions involving positive and negative indices and be able to work with fractional indices and their root equivalents. There are some important results that come from working with indices which can help you understand further work on exponentials and logarithms.

Important rules, results and examples involving indices are given in this table.

Rule	Explanation	Example(s)
$a^0 = 1$	Any (non-zero) value raised to the power zero is equal to 1. The exception here is 0^0 which is undefined.	$7^0 = 1$
$a^1 = a$	The first power of any value is simply that value.	$3^1 = 3$
$a^m \times a^n = a^{m+n}$	Providing they have the same **base** (a in this example), terms with indices can be multiplied and written as a single term where the indices are added.	$x^2 \times x^5 = x^7$ $3h^4 \times 2h^5 = 6h^9$
$a^m \div a^n = a^{m-n}$	Providing they have the same base, terms with indices can be divided and written as a single term where the indices are subtracted.	$t^6 \div t^2 = t^4$ $12p^7 \div 3p^5 = 4p^2$
$(a^m)^n = a^m$	Terms with indices which are then raised to a further power can be simplified and written as a single term where the indices are multiplied.	$(q^2)^5 = q^{10}$ $(5b^3)^2 = 25b^6$
$(\sqrt{a})^2 = \sqrt{a} \times \sqrt{a} = a$	Squaring a surd cancels out the root leaving just the variable or number.	$(\sqrt{2})^2 = 2$
$a^{\frac{m}{n}} = \sqrt[n]{a^m}$	A term with a fractional **index** can be written in root form and vice-versa. This is a crucial skill required for differentiation and integration. The denominator of the fraction becomes the root and the numerator becomes the power.	$k^{\frac{2}{3}} = \sqrt[3]{k^2}$ $8^{\frac{5}{3}} = \sqrt[3]{8^5} =$ $\sqrt[3]{8}^5 = 2^5 = 32$
$a^{-m} = \frac{1}{a^m}$	A variable with a negative index can be written as a fraction where the variable appears on the denominator with a positive index. You must be confident using this relationship. If not, you won't be able to complete the algebra stages at the beginning of many calculus questions. Remember, a negative index does not mean a negative value.	$x^{-4} = \frac{1}{x^4}$ $3y^{-5} = \frac{3}{y^5}$ $\frac{3a^{-2}}{5} = \frac{3}{5a^2}$

Geometry skills

You should know how to find and work with alternate, corresponding and vertically opposite angles and should know angle properties of triangles and common quadrilaterals, e.g. square, rectangle, rhombus, kite, parallelogram. You should also know the angle and symmetry properties of circles. Any of these skills could be required when answering questions involving straight lines, circles, **vectors** or **optimisation** (calculus).

Properties of quadrilaterals

Angles in any quadrilateral (four-sided shape) add up to 360°. Some important properties of common quadrilaterals are given in this table.

Quadrilateral	All angles 90°	Diagonals bisect (cut in half)	Diagonals are perpendicular (meet at 90°)	Opposite sides parallel
Square	✓	✓	✓	✓
Rectangle	✓	✓		✓
Rhombus		✓	✓	✓
Parallelogram		✓		✓
Kite			✓	

Properties of circles

Knowledge of the basic properties of circles is assumed and may be required for questions involving straight lines and further circle work.

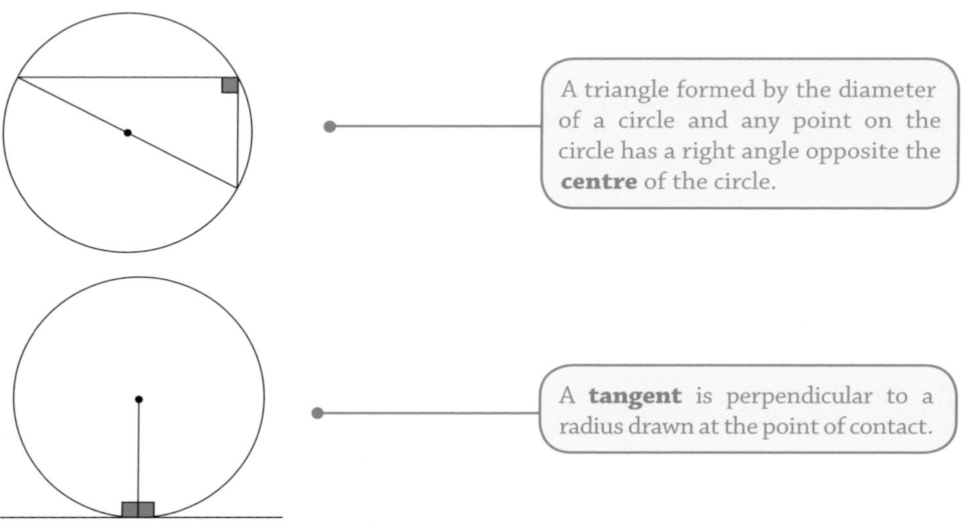

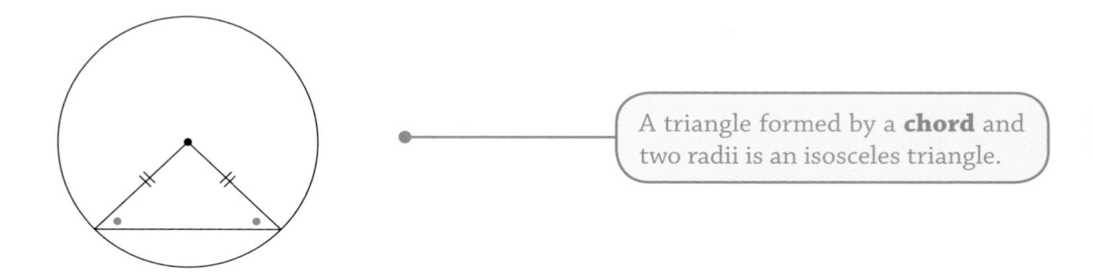

Trigonometry

The National 5 Maths course provides an introduction to trigonometry and you should be familiar with the basic graphs of $y = \sin x$, $y = \cos x$ and $y = \tan x$. In addition to the basic graphs, you should be able to work with trig graphs that have been transformed in some way. You should also have an understanding of how angles greater than 90° can be related to angles less than 90°. You should be able to use your knowledge of related angles and the graphs of trig functions to help you solve trig equations and other problems in context.

Graphs of trigonometric functions

The basic trig graphs are shown below.

The **amplitude** of a trig graph is found using $a = \dfrac{\max - \min}{2}$.

The graphs of $y = \sin x$ and $y = \cos x$ both have an amplitude of 1.

The **frequency** of a trig graph is the number of complete waves between 0 and 360°. The **period** of a trig graph is the length of one complete wave and is found using $\dfrac{360^\circ}{\text{frequency}}$.

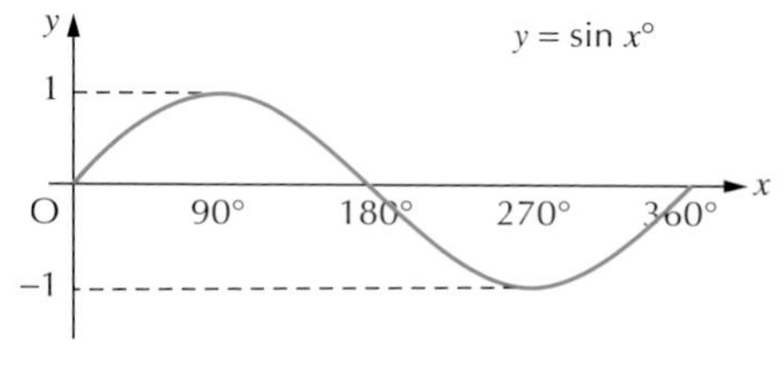

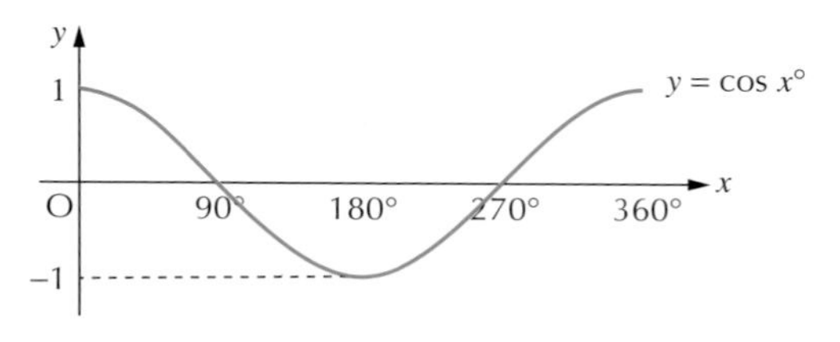

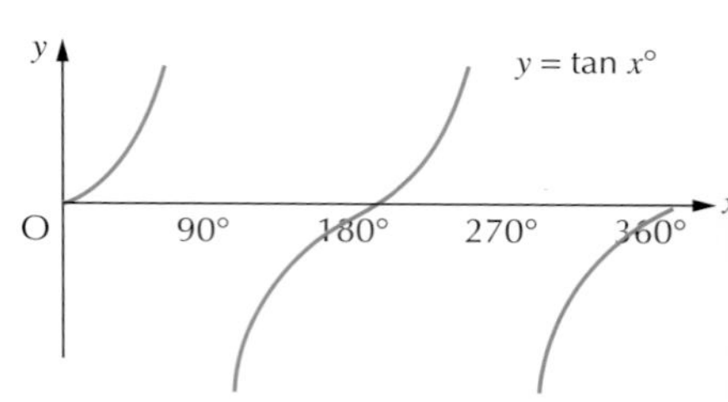

You should have an understanding of the relationship between the transformed trig graphs in this table and their functions.

Function	Effect	Graph
$y = a\sin x$	Changing the value of a affects the amplitude of the graph. $a > 1$ will make the graph 'taller'. $a < 0$ will cause the graph to 'flip' in the x-axis.	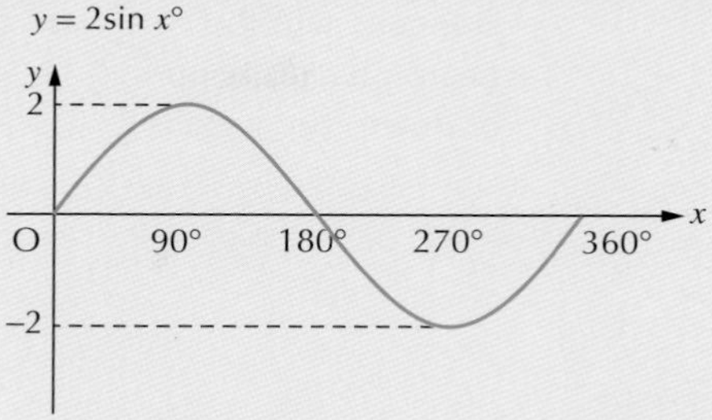 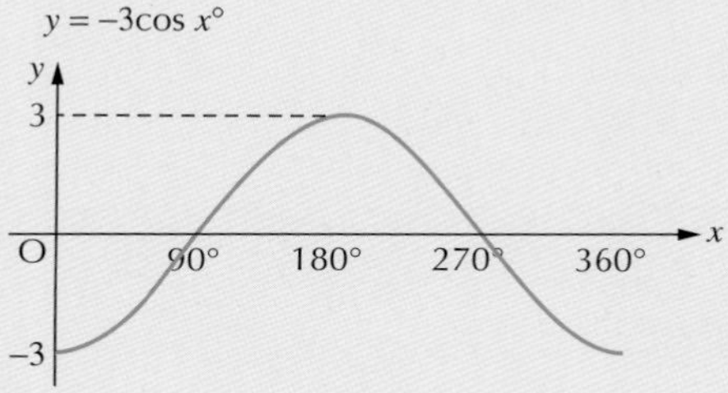 $y = 2\sin x°$ $y = -3\cos x°$
$y = \cos bx$	Changing the value of b affects the frequency of the graph. In fact, there will be 'b' waves between 0 and 360°.	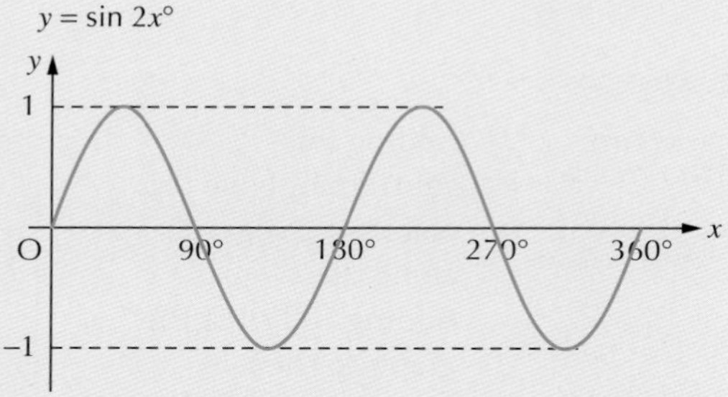 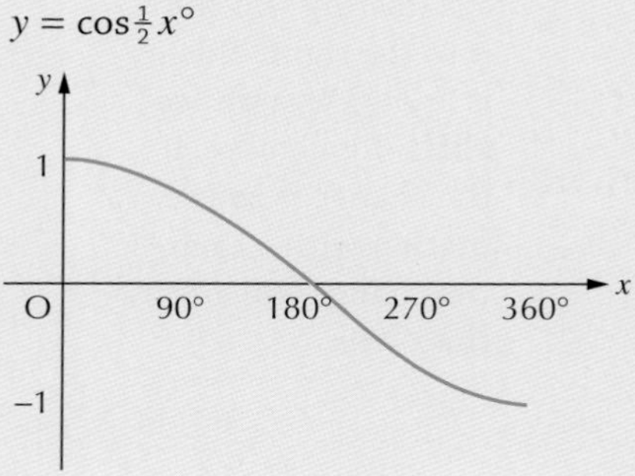 $y = \sin 2x°$ $y = \cos \frac{1}{2}x°$

$y = \sin x \pm c$	Adding (or subtracting) a **constant** will shift the graph c units up (or down) the y-axis.	$y = \sin x° + 3$ 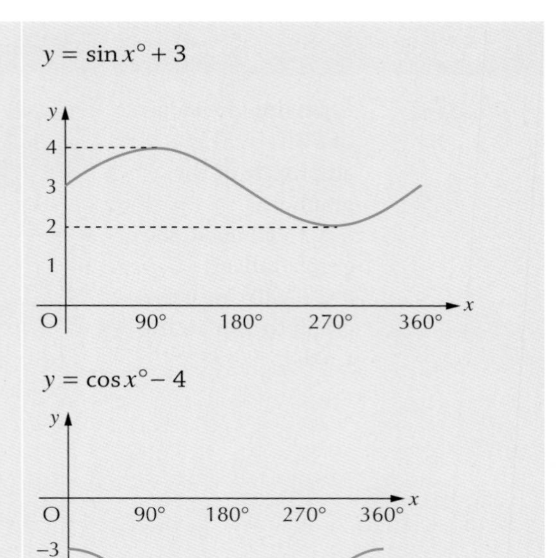 $y = \cos x° - 4$
$y = \sin(x \pm d)$	Adding (or subtracting) like this will shift the graph along the x – axis. Adding will shift the graph to the left, subtracting will shift it to the right. This is known as a **phase shift**. Remember, if the change is **in** the bracket with x then the change will affect the x-axis.	$y = \sin(x - 30)°$ 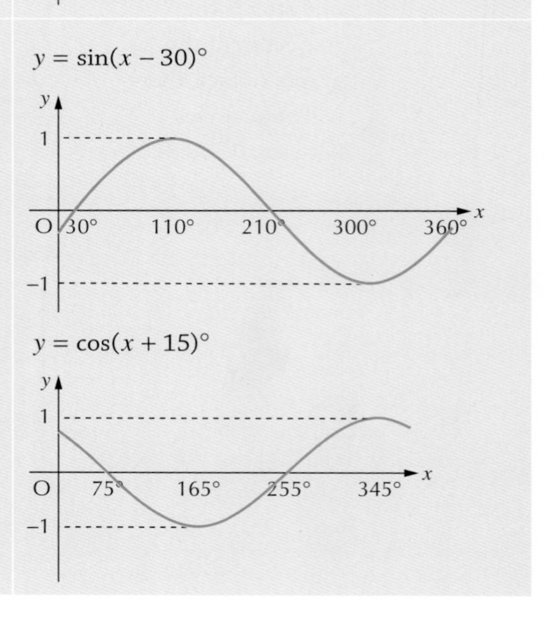 $y = \cos(x + 15)°$

Related angles

Looking at the symmetry in the trig graphs lets us see that, for $0 \leq x \leq 360°$, i.e. values of x between 0 and 360°, there are often two x values where the curve is at the same height. For example, on the graph of $y = \sin x°$, there are two x values with a corresponding y-value of 0.5. These are $x = 30°$ and $x = 150°$. There are also two further values of x for which the corresponding y values are –0.5. These are $x = 210°$ and $x = 330°$.

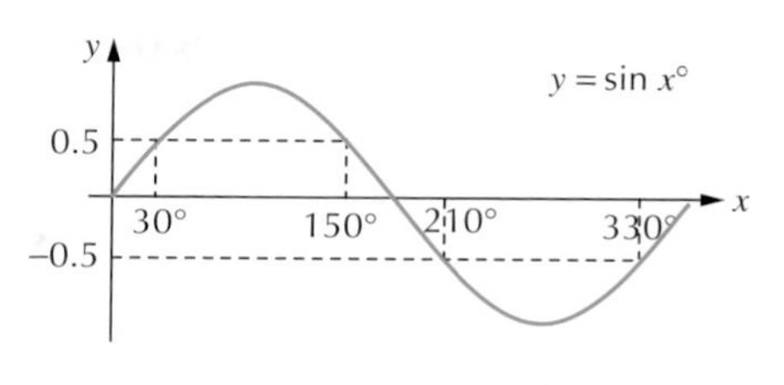

An understanding of this symmetry in the trig graphs allows us to relate angles greater than 90° to angles less than 90°.

$$\sin 150° = \sin 30° = \frac{1}{2}$$

$$\sin 330° = -\sin 30° = -\frac{1}{2}$$

$$\sin 210° = -\sin 30° = -\frac{1}{2}$$

Solving trigonometric equations

Being able to solve trig equations is a vital skill for Higher Maths. Take care with equations that reduce down to a negative **ratio** and remember to use the four-quadrant diagram to help you find all the required solutions.

Example 6.22

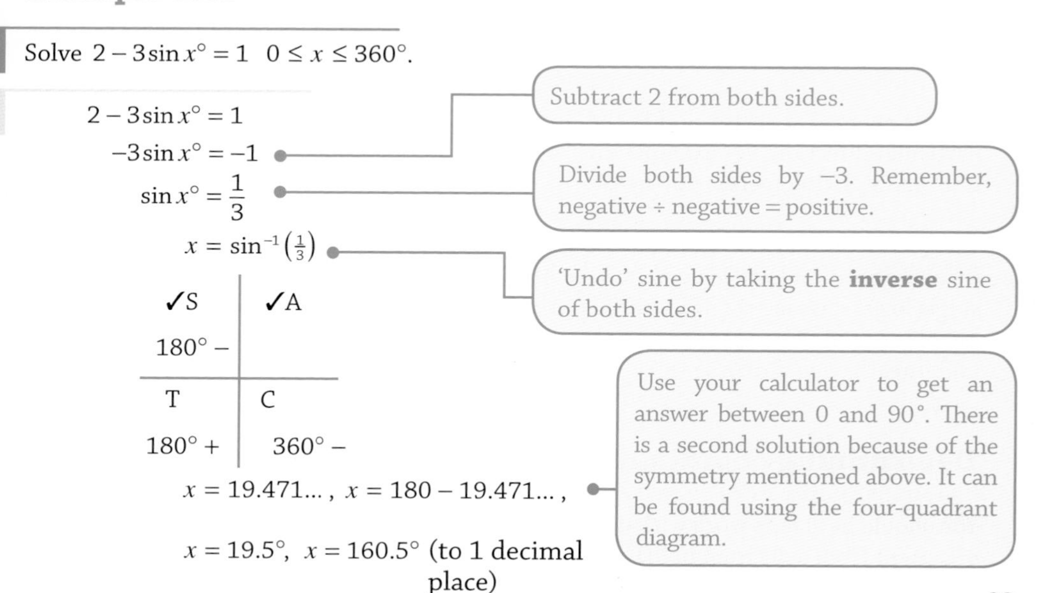

Example 6.23

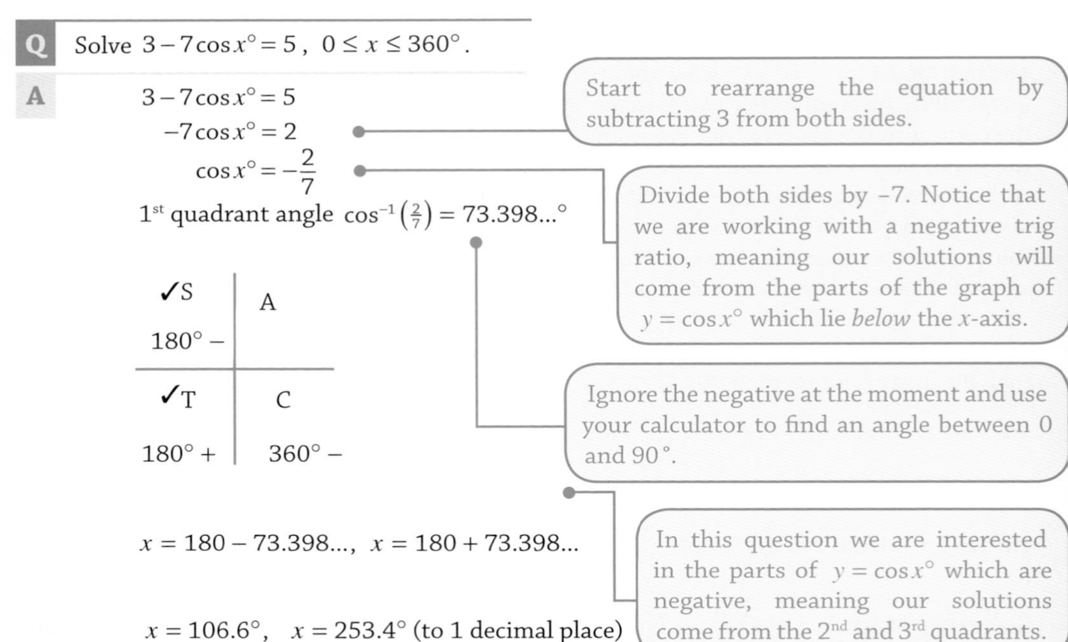

Vectors

Vectors are part of the National 5 Mathematics course and working with vectors in three dimensions (3D) is further developed in the Higher course. A vector is a quantity that has a **magnitude** (size) **and** a direction. Some vector quantities include velocity, acceleration and force. Quantities that have a magnitude but no directional 'sense' are called **scalars**. Examples of scalars include: distance, speed, time, volume and temperature.

Displacement is a vector and is the distance an object has moved in a particular direction. For example the diagram shows the displacement of B from A is 25 kilometres northeast.

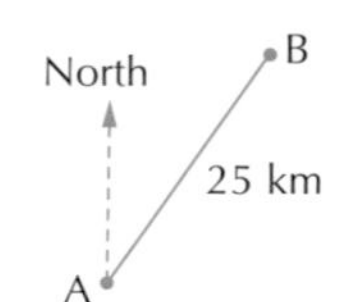

The vector or 'journey' from A to B can be represented by a **directed line segment** and would be written as $\overrightarrow{AB}$. Vectors can also be written as single lower-case letters. If a vector is written in a textbook (or an exam) it will be written in bold, e.g. '**a**'. If you are writing a vector using a single letter it is common to put a line underneath, e.g. '$\underset{\sim}{a}$' or '$\underline{a}$'.

Determining coordinates in three dimensions

You should be able to work with diagrams in 3D and be able to determine the coordinates of a point from a diagram in 3D.

The diagram shows a cube with side length 5 units. Point E is at the origin.

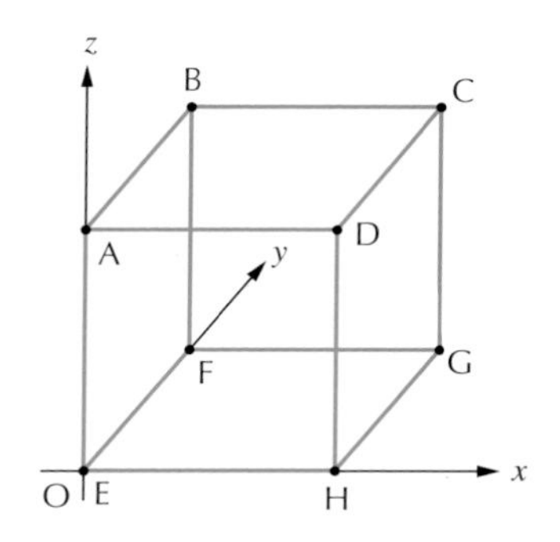

We can use the diagram to determine the coordinates of all the vertices of this cube. For example, point H has coordinates $(5,0,0)$ because it is five units along the x-axis, zero units along the y-axis and zero units along the z-axis. In a similar way we can determine the coordinates of $A(0,0,5)$, $B(0,5,5)$, $C(5,5,5)$ etc.

Coordinates are basically instructions telling us how to get from the origin to a given point. We can represent the journey from the origin to any point as a column vector, e.g. $\overrightarrow{OB} = \mathbf{b} = \begin{pmatrix} 0 \\ 5 \\ 5 \end{pmatrix}$. Vectors starting at the origin are called **position vectors**. Position vectors can be used to determine the 'journey' or vector between two points, e.g. the vector $\overrightarrow{AC} = \mathbf{c} - \mathbf{a} = \begin{pmatrix} 5 \\ 5 \\ 5 \end{pmatrix} - \begin{pmatrix} 0 \\ 0 \\ 5 \end{pmatrix} = \begin{pmatrix} 5 \\ 5 \\ 0 \end{pmatrix}$.

We can go on to find the magnitude of the vector $\overrightarrow{AC}$ using a 3D version of Pythagoras' theorem, e.g. $|\overrightarrow{AC}| = \sqrt{5^2 + 5^2 + 0^2} = \sqrt{50} = 5\sqrt{2}$ units.

Resultant vectors

Two or more vectors can be added (or subtracted) to give a resultant vector.

In the diagram below we can find the resultant vector $\overrightarrow{PT}$ by adding

$$\overrightarrow{PS} + \overrightarrow{SR} + \overrightarrow{RT} = \begin{pmatrix} 6 \\ 0 \\ 0 \end{pmatrix} + \begin{pmatrix} 0 \\ 6 \\ 0 \end{pmatrix} + \begin{pmatrix} -3 \\ -3 \\ 6 \end{pmatrix} = \begin{pmatrix} 3 \\ 3 \\ 6 \end{pmatrix}.$$

We get the same result for $\overrightarrow{PT}$ if we use position vectors, i.e.

$$\overrightarrow{PT} = \mathbf{t} - \mathbf{p} = \begin{pmatrix} 4 \\ 5 \\ 10 \end{pmatrix} - \begin{pmatrix} 1 \\ 2 \\ 4 \end{pmatrix} = \begin{pmatrix} 3 \\ 3 \\ 6 \end{pmatrix}$$

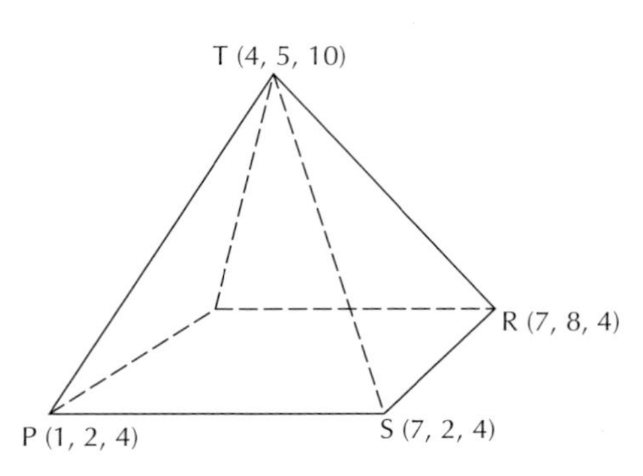

Formulae

Formulae given in the examination

You will be given a list of formulae in your examination. This is usually printed in the inside front cover of your examination paper. You must be confident using all the formulae on the sheet. It is very likely that you will be required to use each of the formulae at some point during your examination. You may also be expected to work with unfamiliar formulae you are unlikely to have seen before. When this happens the formula will be given to you in the text of the question.

Formula list

Circle:

The equation $x^2 + y^2 + 2gf + 2fy + c = 0$ represents a circle with **centre** $(-g,-f)$ and radius $\sqrt{g^2 + f^2 - c}$.

The equation $(x-a)^2 + (y-b)^2 = r^2$ represents a circle with centre (a,b) and radius r.

Scalar product: $\mathbf{a.b} = |a||b|\cos\theta$ where θ is the angle between **a** and **b**

or $\mathbf{a.b} = a_1b_1 + a_2b_2 + a_3b_3$ where $\mathbf{a} = \begin{pmatrix} a_1 \\ a_2 \\ a_3 \end{pmatrix}$ and $\mathbf{b} = \begin{pmatrix} b_1 \\ b_2 \\ b_3 \end{pmatrix}$.

Trigonometric formulae:

$$\sin(A \pm B) = \sin A \cos B \pm \cos A \sin B$$

$$\cos(A \pm B) = \cos A \cos B \mp \sin A \sin B$$

$$\sin 2A = 2\sin A \cos A$$

$$\cos 2A = \cos^2 A - \sin^2 A$$

$$= 2\cos^2 A - 1$$

$$= 1 - 2\sin^2 A$$

Hint

More often than not you will have to use the **scalar product** to find the angle between two **vectors**. In this case you will need to use this re-arranged version of the formula above: $\cos\theta = \dfrac{\mathbf{a.b}}{|\mathbf{a}||\mathbf{b}|}$.

Table of standard derivatives:

$f(x)$	$f'(x)$
$\sin ax$ $\cos ax$	$a\cos ax$ $-a\sin ax$

Table of standard integrals:

$f(x)$	$\int f(x)dx$
$\sin ax$ $\cos ax$	$-\frac{1}{a}\cos ax + C$ $\frac{1}{a}\sin ax + C$

Hint

We would suggest that you use this diagram instead of the tables given above. It's simple and allows you to differentiate and **integrate** −sin and −cos easily.

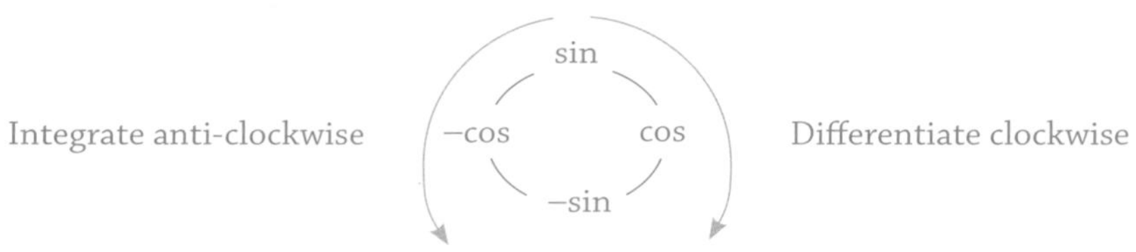

Going clockwise gives **derivatives**, e.g. the derivative of sin is cos. Going anti-clockwise gives integrals, e.g. integrating −sin gives cos. Draw this diagram onto your formula sheet at the start of your examination so you don't forget it.

Useful formulae not given in the examination

It is expected that you will have knowledge of general mathematics below Higher level. This means that you will be expected to complete processes, for example dividing fractions, which are not strictly part of the Higher Maths course. This also means that you should have knowledge of commonly encountered formulae. Included here are some formulae it would be useful to be familiar with before your Higher Maths exam. Although several formulae are included here it should be noted that this list is a guide only. You might also be expected to work with formulae you may not have encountered before, for example the formula for the curved surface area of a hemisphere appears in the SQA specimen paper. In situations like this the unfamiliar formula will be provided for you in the text of the questions.

Area of a circle	$A = \pi r^2$
Area of a rectangle	$A = lb$
Area of a trapezium	$A = \frac{1}{2}h(a+b)$ (Where a and b represent the lengths of the **parallel** sides and h is the perpendicular distance between them.)
Area of a triangle	$A = \frac{1}{2}bh$ (half base × height) or $A = \frac{1}{2}ab\sin C$
Circumference of a circle	$C = \pi d$ or $C = 2\pi r$
Cosine rule	$a^2 = b^2 + c^2 - 2bc\cos A$ or $\cos A = \frac{b^2 + c^2 - a^2}{2bc}$
Discriminant	$b^2 - 4ac$
Distance between points (x_1, y_1) and (x_2, y_2)	$d = \sqrt{(x_2 - x_1)^2 + (y_2 - y_1)^2}$
Gradient	$m = \frac{y_2 - y_1}{x_2 - x_1}$ or $m = \tan\theta$ (Where θ is the angle a line makes with the positive direction of the x – axis.)
Limit (where one exists) of a sequence	$L = \frac{b}{1-a}$ (see chapter 18 for more on sequences)
Midpoint between points (x_1, y_1) and (x_2, y_2)	$M = \left(\frac{x_1 + x_2}{2}, \frac{y_1 + y_2}{2}\right)$
Pythagoras' theorem	$c^2 = a^2 + b^2$ or $c = \sqrt{a^2 + b^2}$

Quadratic formula	$x = \frac{-b \pm \sqrt{b^2 - 4ac}}{2a}$
Sine rule	$\frac{a}{\sin A} = \frac{b}{\sin B} = \frac{c}{\sin C}$
Straight line	$y - b = m(x - a)$
Surface area of a closed cylinder	$A = 2\pi r^2 + 2\pi rh$
Surface area of a cube	$A = 6l^2$
Surface area of a cuboid	$A = 2lb + 2bh + 2lh$
Surface area of an open cylinder	$A = 2\pi rh$
Trigonometric identities	$\sin^2 A + \cos^2 A = 1$ and $\frac{\sin A}{\cos A} = \tan A$
Volume of a cone	$V = \frac{1}{3}\pi r^2 h$
Volume of a cuboid	$V = lbh$
Volume of a cylinder	$V = \pi r^2 h$
Volume of any prism	$V = Ah$ (Where A is the area of the base of the prism.)
Volume of any pyramid	$V = \frac{1}{3}Ah$ (Where A is the area of the base of the pyramid.)
Volume of a sphere	$V = \frac{4}{3}\pi r^3$

Unit 1: Expressions and functions

Exponential and logarithmic functions

Before you start this chapter you should be able to:

- Use the laws of indices/**exponents** to simplify **expressions** (Chapter 6).

This chapter covers:

- Simplifying a numerical expression using the laws of **logarithms** and exponents
- Solving logarithmic and **exponential** equations
- Working with relationships of the form $y = ax^b$ and $y = ab^x$
- Using maths to model situations involving the logarithmic or exponential **function**.

The laws of logarithms

Logarithms and exponentials can seem quite daunting at first but it's worth remembering that there are set rules for working with them. Learn the rules and apply them in the correct order and you will see everything fall into place. Knowledge of the laws of indices (also known as exponents) is essential, see Chapter 6 (Essential mathematical skills) for more on working with indices.

Important rules, results and examples involving logarithms are given in this table.

Rule	Explanation	Example(s)
$\log_a 1 = 0$	• The logarithm of 1 to any **base** equals 0.	$\log_4 1 = 0$
$\log_a a = 1$	The logarithm of any number to its own base equals 1.	$\log_9 9 = 1$

$\log_a p + \log_a q = \log_a (pq)$	To add two logarithmic terms with the same base, multiply the arguments (p and q) and write as a single logarithmic term.	$\log_a 3 + \log_a 4 = \log_a 12$ $\log_2 x + \log_2 (x+1) = \log_2 (x^2 + x)$
$\log_a x - \log_a y = \log_a \left(\frac{x}{y}\right)$	To subtract two logarithmic terms with the same base, divide the arguments and write as a single logarithmic term.	$\log_a 12 - \log_a 4 = \log_a 3$
$\log_a (m^n) = n\log_a m$	Logarithms with exponents can be manipulated using this law.	$\log_a a^2 = 2\log_a a = 2(1) = 2$ $3\log_8 2 = \log_8 2^3 = \log_8 8 = 1$

Simplify a numerical expression using the laws of logarithms and exponents

Before evaluating or solving a logarithmic expression or equation you will usually have to use the laws of logarithms to simplify the expression or equation first. Knowing the first few powers of numbers like 2, 3 and 4 will help when evaluating logarithmic and exponential expressions.

Example 8.1

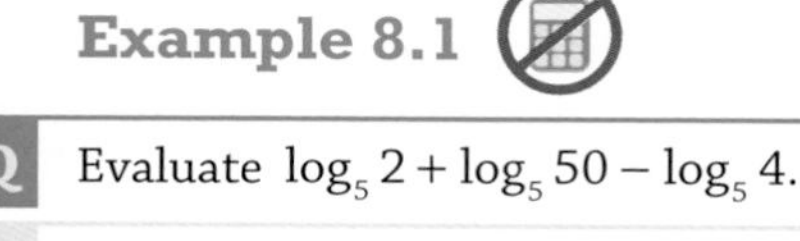

Q Evaluate $\log_5 2 + \log_5 50 - \log_5 4$.

A

$\log_5 2 + \log_5 50 - \log_5 4$

$= \log_5 100 - \log_5 4$ (✓) — Work from left to right and combine the first two terms by multiplying the '2' by the '50'.

$= \log_5 25$ (✓)

$= \log_5 5^2$ — Writing the argument, i.e. the '25' as a power of the base will make the final step easier.

$= 2\log_5 5$ — Use log laws to turn the exponent into a multiplier.

$= 2$ (✓) — Since $\log_5 5 = 1$ our answer simplifies to a single number term.

3

Keyword

The word 'evaluate' tells you that your final answer should be a number.

Solve logarithmic and exponential equations

There are lots of marks available to you in these questions; you just need to set out on the right foot! Being able to remove a logarithmic expression and knowing how this affects the rest of an equation is a vital skill.

Example 8.2

Q 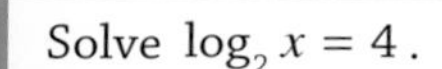
Solve $\log_2 x = 4$. **2**

A

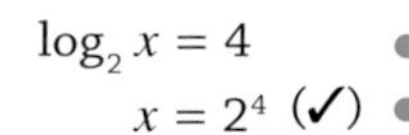

$$\log_2 x = 4$$

Here we have a logarithmic expression on the LHS of the equation and a 'normal' number on the RHS.

$$x = 2^4 \ (✓)$$

The logarithm on the LHS is removed with an exponential operation. This has the effect of turning the 4 into an exponent on the RHS.

$$x = 16 \ (✓)$$

Use your knowledge of the powers of 2 to simplify your answer. See Chapter 6 (Essential mathematical skills) for more on commonly used powers and roots.

Example 8.3

Q 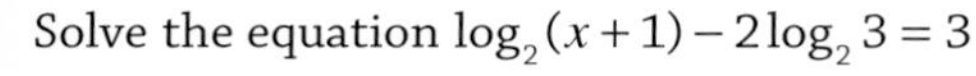
Solve the equation $\log_2(x+1) - 2\log_2 3 = 3$ **4**

A

$$\log_2(x+1) - 2\log_2 3 = 3$$

$$\log_2(x+1) - \log_2 3^2 = 3 \ (✓)$$

Use the laws of logarithms to write multipliers as exponents.

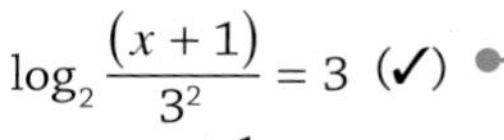

$$\log_2 \frac{(x+1)}{3^2} = 3 \ (✓)$$

Use log laws to write the LHS as a single logarithmic term.

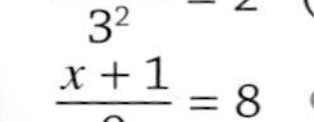

$$\frac{x+1}{3^2} = 2^3 \ (✓)$$

'Removing' the logarithm turns the 3 on the RHS into an exponent. Notice that the base remains the same, i.e. 2.

$$\frac{x+1}{9} = 8$$

Simplify.

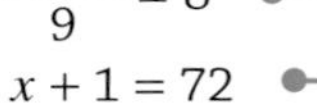

$$x + 1 = 72$$

Multiply both sides by 9.

$$x = 71 \ (✓)$$

Hint

As with most equations, a good strategy is to collect the terms with x together on the left-hand side and collect all the other (number) terms together on the right-hand side.

Example 8.4

Q Find x if $2\log_x 4 - 5\log_x 2 = 1$. **4**

A

$$2\log_x 4 - 5\log_x 2 = 1$$

$$\log_x 4^2 - \log_x 2^5 = 1 \ (\checkmark)$$

Use the log laws to write the 2 and the 5 as exponents.

$$\log_x \frac{4^2}{2^5} = 1 \ (\checkmark)$$

Using the log laws to write as a single log term will get you the second mark.

$$\log_x \frac{16}{32} = 1$$

This is where it helps to know the first few powers of numbers like 2 and 4.

$$\log_x \frac{1}{2} = 1 \ (\checkmark)$$

Simplify.

$$\frac{1}{2} = x^1$$

'Removing' the log turns the 1 on the RHS into an exponential. Notice that the base is still the same, i.e. x.

$$x = \frac{1}{2} \ (\checkmark)$$

Working with relationships of the form $y = ax^b$ and $y = ab^x$

These questions are quite often set in context, for example the concept of 'half-life'. Half-life is often related to radioactive decay, although a more general definition of half-life is 'the time taken for a specific property to decrease by half'. It might sound obvious but it is important to remember that t represents time and at the very start of this type of situation t will be equal to zero.

Example 8.5

Q The concentration of the pesticide, *Xpesto,* in soil can be modelled by the equation.

$$P_t = P_0 e^{-kt}$$

where:

- P_0 is the initial concentration;
- P_t is the concentration at time t;
- t is the time, in days, after the application of the pesticide.

Once in the soil, the half-life of a pesticide is the time taken for its concentration to be reduced to one half of its initial value.

(a) If the half-life of *Xpesto* is 25 days, find the value of k to 2 significant figures. **4**

On all *Xpesto* packaging, the manufacturer states that 80 days after application the concentration of *Xpesto* in the soil will have decreased by over 90%.

(b) Is this statement correct? Justify your answer. **4**

A (a)

$P_t = P_0 e^{-kt}$

$P_t = \frac{1}{2}P_0$ — After one 'half-life' the amount of pesticide remaining (P_t) will be half the original amount, i.e. $\frac{1}{2}P_0$.

$P_0 e^{-25k} = \frac{1}{2}P_0$ (✓) — Equate the expressions for P_t; interpreting 'half-life' in this way secures the first mark.

$e^{-25k} = \frac{1}{2}$ (✓) — Simplify by dividing both sides by P_0 to get the second mark.

$\ln e^{-25k} = \ln\left(\frac{1}{2}\right)$ — Take the **natural logarithm** of both sides. Remember, $\ln\left(\frac{1}{2}\right)$ is the same as writing $\log_e\left(\frac{1}{2}\right)$.

$-25k \ln e = \ln\left(\frac{1}{2}\right)$ — Use log laws to write the exponent, i.e. $-25k$, as a multiplier.

$-25k = \ln\left(\frac{1}{2}\right)$ (✓) — Simplify using the fact that $\ln e = 1$.

$k = \frac{\ln\left(\frac{1}{2}\right)}{-25}$ — Divide both sides by −25. You can key this into your calculator in one go, just take care and remember to put brackets around the $\frac{1}{2}$.

$k = 0.027725...$

$k \approx 0.028$ to 2 significant figures (✓) — Make sure you round your answer as instructed in the question.

Hint

This question is worth a lot of marks and there is quite a bit of information to read. However, the context doesn't affect how you should approach the question and once you have practiced a few like these, you should start to realise that they are almost always answered in the same way.

(b)

$P_t = P_0 e^{-kt}$

$P_t = P_0 e^{-0.028(80)}$ (✓)

$P_t = 0.10645...P_0$ (✓)

$1 - 0.10645... = 0.8935...$

$0.8935... \approx 89.35\%$ (✓)

No, the statement is not correct as the concentration of pesticide will only have decreased by 89.35 % after 80 days and $89.35 < 90$. (✓)

Relate the answers to your calculations back to the context of the question and always include a numerical comparison to make sure you get the final mark.

Hint

This question is essentially saying, 'Use your answer to (a) to find out how much remains after 80 days.' Just make sure you relate your answer back to the 90% mentioned in the question.

Substitute $t = 80$ and the value of k from part (a) into the equation.

Use your calculator to work out $e^{-0.028(80)} = 0.10645...$. This is a decimal representation of how much pesticide *remains* after 80 days.

Subtract from 1 (one whole) to find out how much pesticide has been *lost*. Keep the answers to each stage on your calculator to avoid any rounding errors.

Manipulating trigonometric expressions

Before you start this chapter you should be able to:

- Identify related angles (Chapter 6)
- Use **exact value ratios** for the sine, cosine and **tangent** of 30°, 45°, 60° and multiples of 90° (Chapter 6)
- Prove trigonometric identities using: $\sin^2 A + \cos^2 A = 1$ and $\dfrac{\sin A}{\cos A} = \tan A$
- Sketch and identify graphs of the form $y = k\cos(x \pm \alpha)$ or the sine equivalent (Chapter 6)
- Use the term **phase angle**.

This chapter covers:

- Converting between **radians** and degrees, e.g. $\dfrac{\pi}{6}$ radians $= 30°$ etc.
- Finding exact values for trigonometric ratios, e.g. $\sin\dfrac{\pi}{4} = \dfrac{1}{\sqrt{2}}$ etc.
- Applying the addition formulae
- Applying the double angle formulae
- Applying trigonometric identities
- Converting $a\cos x + b\sin x$ to $k\cos(x \pm \alpha)$ or $k\sin(x \pm \alpha)$, $k > 0$.

Converting between radians and degrees

Degrees are not the only unit used to measure the size of angles. It is essential that you recognise when you should use degrees and when you should use radians. Luckily it is simple: if the question has a degree symbol in it, e.g. $\sin x° = \frac{1}{2}$ then you should work in degrees, if not, use radians. To make extra certain, trigonometric equations will have a fixed **domain**, e.g. $0 \le x \le 360$ or $0 \le x \le 2\pi$. If the domain contains a π symbol then you should work in radians, if not, you should work in degrees. There is a symbol for radians $(^c)$ but it is not commonly used and it has become normal for an angle in radians to simply be denoted by a number, e.g. $\frac{3\pi}{2}$, although in some contexts you might see $\frac{3\pi}{2}$ rad or $\frac{3\pi}{2}$ radians.

It is important to remember that **differentiation** and **integration** involving trigonometric **functions** simply does not work if you use degrees so being confident with radians is really important.

Some people prefer to convert from radians to degrees before tackling a question. There is no problem with doing this unless you lose a mark because you forget to convert the final answer back to radians. Most scientific calculators can convert between degrees and radians but you might need to search on the internet for an instruction manual to tell you how! Once you have identified that a question requires radians, it is a good idea to work with radians throughout. If you do want to convert between degrees and radians, the following diagram and examples will show you how.

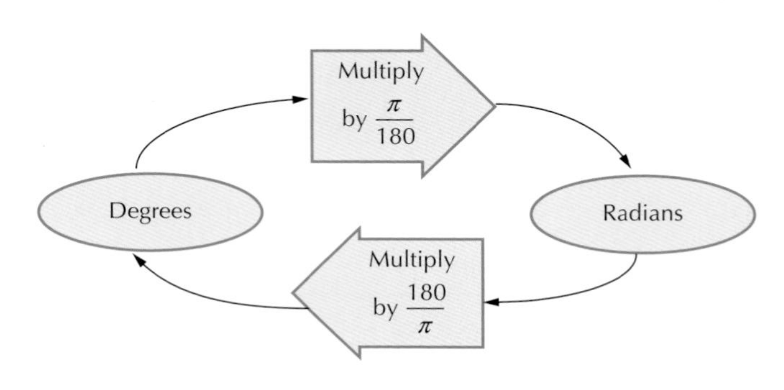

Example 9.1

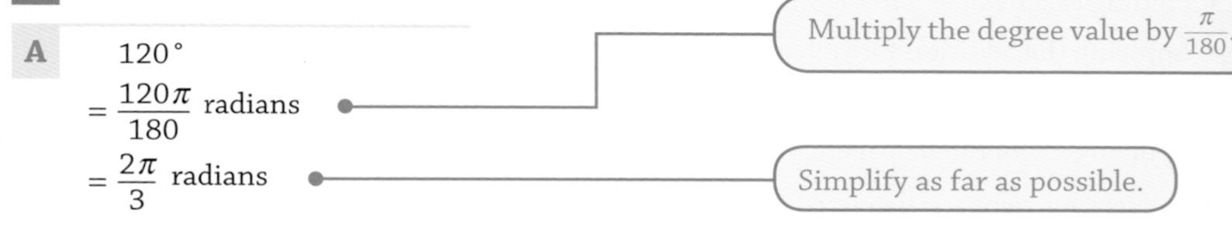

Q Convert $120°$ to radians.

A

$$120°$$
$$= \frac{120\pi}{180} \text{ radians}$$
$$= \frac{2\pi}{3} \text{ radians}$$

Example 9.2

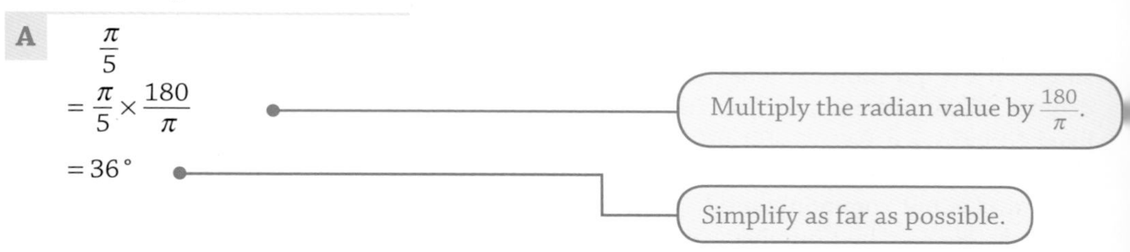

Q Convert $\frac{\pi}{5}$ radians to degrees.

A

$$\frac{\pi}{5}$$
$$= \frac{\pi}{5} \times \frac{180}{\pi}$$
$$= 36°$$

Finding exact values for trigonometric ratios

There are some angles for which there are exact values for sine, cosine and tangent. Exact values are often fractions or surds. For example, the exact value of $\sin 45°$ is $\frac{1}{\sqrt{2}}$. Working with exact values avoids having to use decimal approximations and is something you must be confident with before your examination. One helpful way to find some commonly used exact value ratios (for angles measured in degrees or radians) is to remember the two triangles shown below.

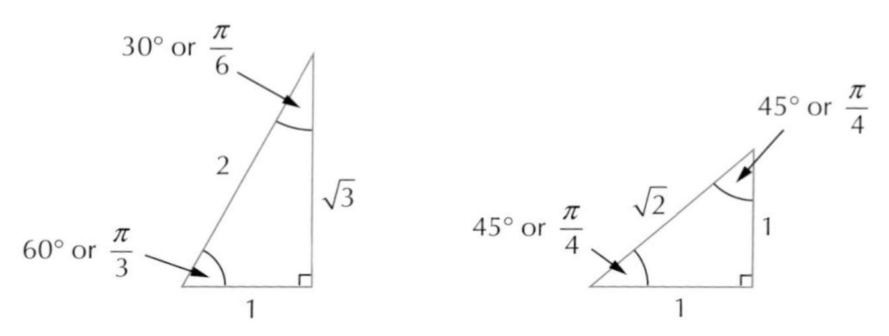

You might find it helpful to draw these triangles on your formula sheet at the start of your examination and refer to them when needed.

If you want to know the exact value of $\cos \frac{\pi}{6}$, find $\frac{\pi}{6}$ and identify the side of the triangle that is adjacent to $\frac{\pi}{6}$, using the fact that cosine is $\frac{\text{adjacent}}{\text{hypotenuse}}$ you can see that $\cos \frac{\pi}{6} = \frac{\sqrt{3}}{2}$.

The rest of the exact values can be read from the triangles in a similar way and these, along with some other important values, are summarised in this table.

Angle (Degrees)	Angle (Radian)	Sine	Cosine	Tangent
0	0	0	1	0
30°	$\frac{\pi}{6}$	$\frac{1}{2}$	$\frac{\sqrt{3}}{2}$	$\frac{1}{\sqrt{3}}$
45°	$\frac{\pi}{4}$	$\frac{1}{\sqrt{2}}$	$\frac{1}{\sqrt{2}}$	1
60°	$\frac{\pi}{3}$	$\frac{\sqrt{3}}{2}$	$\frac{1}{2}$	$\sqrt{3}$
90°	$\frac{\pi}{2}$	1	0	Undefined
180°	π	0	−1	0
270°	$\frac{3\pi}{2}$	−1	0	Undefined
360°	2π	0	1	0

Applying the addition formula and applying exact values for trigonometric ratios

Example 9.3

Q (a) Find an equivalent **expression** for $\cos(x+30)°$. **1**

(b) Hence, or otherwise, determine the exact value of $\cos 75°$. **3**

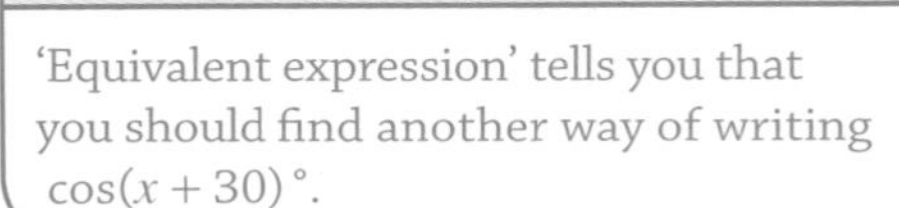

Keyword

'Equivalent expression' tells you that you should find another way of writing $\cos(x+30)°$.

Keyword

The word 'exact' tells you that your answer will probably be a fraction and will often contain a surd. Exact value triangles will help here.

A (a) $\cos(x+30)°$

$= \cos x° \cos 30° - \sin x° \sin 30°$ (✓)

Use the expansion $\cos(A+B) = \cos A \cos B - \sin A \sin B$ from the formula sheet but make sure you replace A and B with the values given in the question.

(b) $\cos 75°$

$= \cos(45+30)°$ (✓)

Part (a) gives a hint that you should use the addition formula in part (b). Set up an addition formula by re-writing 75° as 30° + 45°. We use 30 and 45 here because they are angles that appear on the exact value triangles.

$= \cos 45° \cos 30° - \sin 45° \sin 30°$

Use the expansion $\cos(A+B) = \cos A \cos B - \sin A \sin B$ from part (a) but make sure you replace A and B with 45° and 30°.

$= \left(\frac{1}{\sqrt{2}}\right)\left(\frac{\sqrt{3}}{2}\right) - \left(\frac{1}{\sqrt{2}}\right)\left(\frac{1}{2}\right)$ (✓)

Use the exact value triangles to make substitutions for cos 45°, cos 30°, sin 45° and sin 30°.

$= \frac{\sqrt{3}}{2\sqrt{2}} - \frac{1}{2\sqrt{2}}$

Multiply to create two fractions with the same denominator.

$= \frac{\sqrt{3}-1}{2\sqrt{2}}$ (✓)

Write as a single fraction.

Applying the double angle formula

It is usually straightforward to identify a double angle question because it contains a double angle like $\sin 2x$ or $\cos 2x$. If you can spot this and remember that you have substitutions available for these double angles on your formula sheet, you should have no problem getting good marks in this type of question.

Hint

If you see $\cos^2 x$ or $\sin^2 x$ remember this is a short way of writing $(\cos x)^2$ or $(\sin x)^2$. To avoid any confusion some people find it helps to re-write these expressions in full.

Example 9.4

If $\cos 2x = \frac{7}{25}$ and $0 < x < \frac{\pi}{2}$, find the exact values of $\cos x$ and $\sin x$. **4**

Hint

Don't be put off by the $0 < x < \frac{\pi}{2}$, this isn't part of the equation but does tell you that your answers should be in radians and that they must be between 0 and $\frac{\pi}{2}$.

$\cos 2x = \frac{7}{25}$

$2\cos^2 x - 1 = \frac{7}{25}$ (✓)

Replace $\cos 2x$ with one of the expressions given in your formula sheet.

$2\cos^2 x = \frac{32}{25}$

Add 1 to both sides of the equation. Remember, 1 is the same as $\frac{25}{25}$.

$\cos^2 x = \frac{16}{25}$ (✓)

Divide both sides of the equation by 2. Remember, half of $\frac{32}{25}$ is $\frac{16}{25}$.

$\cos x = \pm\frac{4}{5}$

Square-root both sides. To square root a fraction simply square root the numerator and square root the denominator but remember, there are two answers to every square root, a positive and a negative.

$\cos x = \frac{4}{5}$ (✓)

Using your knowledge of the graph of $y = \cos x$ you can disregard the $-\frac{4}{5}$ in this case because $y = \cos x$ is negative when $\frac{\pi}{2} < x < \frac{3\pi}{2}$, which is outside the domain stated in the question. See Chapter 10 (Functions and graphs) for more on the domain of a function.

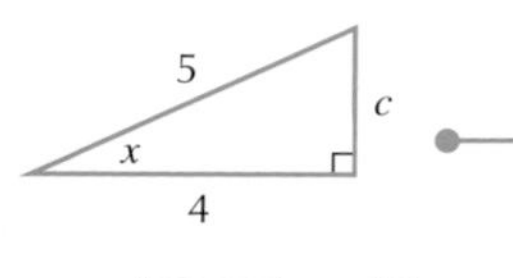

$c = \sqrt{5^2 - 4^2} = \sqrt{9} = 3$

$\sin x = \frac{3}{5}$ (✓)

Remember, $\cos x$ is defined as being $\frac{\text{adjacent}}{\text{hypotenuse}}$ and $\sin x$ is defined as $\frac{\text{opposite}}{\text{hypotenuse}}$. By sketching a right-angled triangle and using Pythagoras' theorem you can find the value of the opposite side.

Converting $a\cos x + b\sin x$ to $k\cos(x \pm \alpha)$ or $k\sin(x \pm \alpha)$, $k > 0$

Commonly known as the wave function, this process allows you to write an expression involving two trigonometric functions as a new expression involving only one trigonometric function. From here you can go on to solve equations and answer questions involving the graph of the function. The angle α has the effect of shifting the graph of a trigonometric function horizontally along the x-axis and is known as the phase angle. $+\alpha$ shifts the graph α units to the left while $-\alpha$ shifts the graph α units to the right.

Example 9.5 

Q (a) The expression $\sqrt{3}\sin x° - \cos x°$ can be written in the form $k\sin(x - a)°$, where $k > 0$ and $0 \le a < 360$. Calculate the values of k and a. **4**

(b) Determine the maximum value of $4 + 5\cos x° - 5\sqrt{3}\sin x°$, where $0 \le x < 360$. **2**

A (a)

$$\sqrt{3}\sin x° - \cos x° = k\sin(x - a)°$$

$$\sqrt{3}\sin x° - \cos x° = k\sin x°\cos a° - k\cos x°\sin a° \quad (✓)$$

$$\sqrt{3}\sin x° - \cos x° = (k\cos a°)\sin x° - (k\sin a°)\cos x°$$

k and a are commonly used in this type of question, although other letters such as r and α are sometimes used.

Use the expansion for $\sin(A - B)$ from the formula sheet but make sure you use the letters in the question, i.e. x and a, if you don't you will lose a mark. Remember, k should appear twice in your expansion!

Rearrange your expansion and compare **coefficients** of $\sin x$ and $\cos x$.

Hint

Showing appropriate working is essential here as marking schemes are very strict for this type of question. Don't take any shortcuts and always fully simplify or you will lose marks. Questions requiring radians are tackled in exactly the same way.

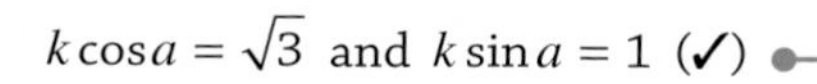

$k\cos a = \sqrt{3}$ and $k\sin a = 1$ (✓)

Make sure you include the k.

$k = \sqrt{\left(\sqrt{3}\right)^2 + 1^2} = \sqrt{4} = 2$ (✓)

Remember $\left(\sqrt{3}\right)^2 = 3$ and make sure you simplify $\sqrt{4}$ to 2.

$\tan a = \frac{1}{\sqrt{3}}$

Take care not to get the ratio upside down! Remember that $\tan a = \frac{k\sin a}{k\cos a} = \frac{1}{\sqrt{3}}$.

✓S	A✓✓
T	C✓

$\sin a$ and $\cos a$ are both positive in this question. Draw a four-quadrant diagram and tick the quadrants where sine is positive, i.e. 1st and 2nd, then tick the quadrants where cosine is positive, i.e. 1st and 3rd. This leaves two ticks in the 1st quadrant and tells us that, in this question, the value of a will come from the first quadrant.

$a = \tan^{-1}\left(\frac{1}{\sqrt{3}}\right)$

$a = 30°$ (✓)

$\sqrt{3}\sin x° - \cos x° = 2\sin(x - 30)°$

(b)

$4 + 5\cos x° - 5\sqrt{3}\sin x°$

Removing a common **factor** of –5 from the trigonometric terms reveals the original expression from part (a).

$= 4 - 5\left(\sqrt{3}\sin x° - \cos x°\right)$

Substitute in your result from part (a).

$= 4 - 5\left(2\sin(x - 30)°\right)$ (✓)

Simplify.

$= 4 - 10\sin(x - 30)°$

Maximum value is 14. (✓)

Use your knowledge of trigonometric graphs: the maximum value that $y = -10\sin(x - 30)°$ reaches is +10. Adding the 4 gives a maximum value of 14 for $y = 4 - 10\sin(x - 30)°$. See Chapter 10 (Functions and graphs) for more on finding the maximum and minimum values of a trig function.

Hint

Part (b) looks complicated but connecting it to your answer to part (a) will make it much more accessible.

Functions and graphs

Before you start this chapter you should be able to:

- Complete the square of a quadratic **function** with a unitary **coefficient** of x^2 and use the completed square form to sketch the graph (Chapter 6)
- Know how to sketch or determine the equation of the graphs of trigonometric functions of the form $y = a\sin bx°$ or $y = a\sin(x + d)°$ and their cosine equivalents (Chapter 6).

This chapter covers:

- Using the terms **domain** and **range**
- Identifying and sketching a function after a translation of the form $af(x)$, $f(bx)$, $f(x) \pm c$, $f(x \pm d)$ or a combination of these
- Sketching the graph of the **derived function** $y = f'(x)$ given the graph of $y = f(x)$
- Sketching the graph of an **exponential** or **logarithmic function** and finding the equation when given the graph
- Sketching the graph or finding the equation of a function of the form $y = a\sin(bx + d) + c$ or $y = a\cos(bx + d) + c$
- Finding the maximum and minimum **values** of a trigonometric function of the form $y = a\sin(bx + d) + c$ or $y = a\cos(bx + d) + c$
- Completing the square where the coefficient of x^2 is non-unitary
- Finding a formula for a **composite function** $f(g(x))$
- Finding a formula for the **inverse** function $f^{-1}(x)$.

Know and use the terms domain and range

The domain of a function is the **set** of input values, i.e. the x values, for which the function is defined, i.e. an output value exists. The range is all the output values of a function, i.e. the y values. The range of a function is easier to figure out if you have a graph of the function. When considering the domain of a function you want to avoid values of x that will cause the function to divide by zero, or to take the square root or logarithm of a negative number. See the additional information in the glossary for details of the different sets which numbers can belong to.

Example 10.1 

A function f is defined by $f(x) = \dfrac{x+1}{2x^2-5x-3}$.

State a suitable domain for f and justify your answer. **2**

$2x^2 - 5x - 3 = 0$

$(2x+1)(x-3) = 0$

$x = -\frac{1}{2}$ or $x = 3$

When $x = -\frac{1}{2}$ or $x = 3$ the function $f(x) = \dfrac{x+1}{0}$, which is undefined. (✓)

Therefore x can be any real number except $-\frac{1}{2}$ or 3 (✓)

Any value of x which creates a zero denominator cannot be in the domain. Such values must be avoided for the function to be properly defined.

Factorise and solve to find the values of x which make the denominator equal to zero. These are the values of x we want to avoid.

Clearly communicate your answer and your reasoning. This answer could also be written as '$x \in \mathbb{R}$, $x \neq -\frac{1}{2}, 3$'. See the glossary for more on this type of notation.

Identify and sketch a function after a translation of the form $af(x)$, $f(bx)$, $f(x) \pm c$, $f(x \pm d)$ or a combination of these

The graph of a function can be moved up, down, left or right, stretched, squashed and reflected. It can be helpful to remember that, if the 'change' is in the brackets with x then the 'change' will affect the x-axis, otherwise the change will affect the y-axis. To get full marks when you draw a transformed function you must mark on the **image** (new position) of all of **all** the points that were given on the original function.

A summary of these changes and the corresponding change to the equation of the function is given in the table below.

Related function	Transformation in words	Transformation in coordinates	Transformation represented graphically
$y = af(x)$ e.g. $y = 3f(x)$	Graph stretched vertically if $a > 1$ and compressed vertically if $0 < a < 1$. Graph also reflected in the x-axis if $a < 0$.	$(x,y) \to (x,ay)$ e.g. $(2,3) \to (2,9)$	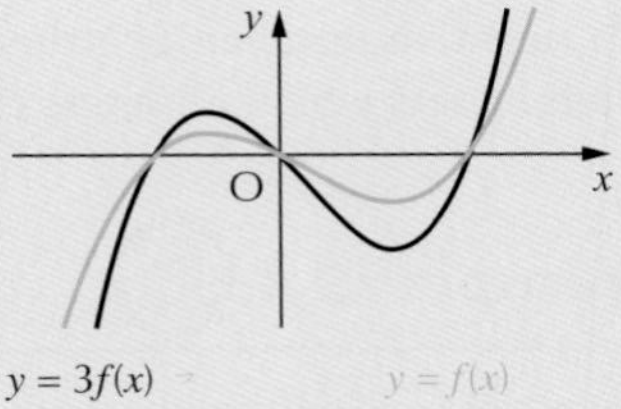 $y = 3f(x)$ $y = f(x)$
$y = f(bx)$ e.g. $y = f(2x)$	Graph compressed horizontally if $b > 1$ and stretched horizontally if $0 < b < 1$. Graph also reflected in the y-axis if $b < 0$.	$(x,y) \to \left(\frac{x}{b},y\right)$ e.g. $(2,3) \to (1,3)$	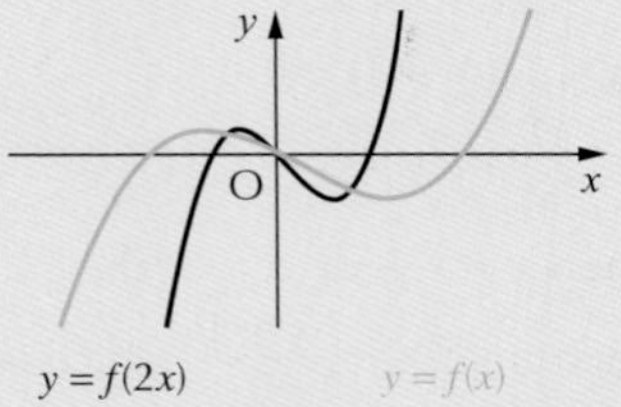 $y = f(2x)$ $y = f(x)$
$y = f(x)+c$ e.g. $y = f(x)+1$	Graph shifted up the y-axis if $c > 0$ and shifted down if $c < 0$.	$(x,y) \to (x,y+c)$ e.g. $(2,3) \to (2,4)$	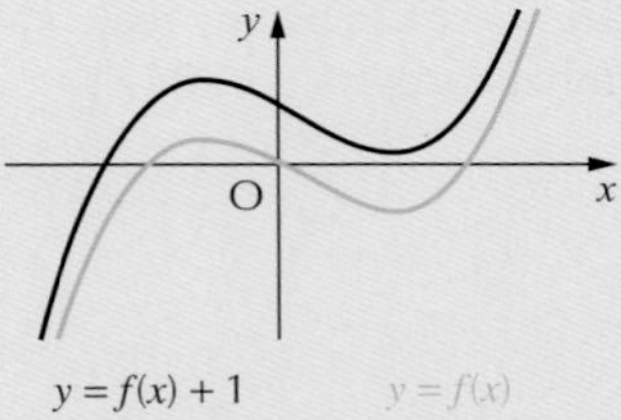 $y = f(x) + 1$ $y = f(x)$
$y = f(x+d)$ e.g. $y = f(x+2)$	Graph shifted left on the x-axis if $d > 0$ and shifted right on the x-axis if $d < 0$.	$(x,y) \to (x-d,y)$ e.g. $(2,3) \to (0,3)$	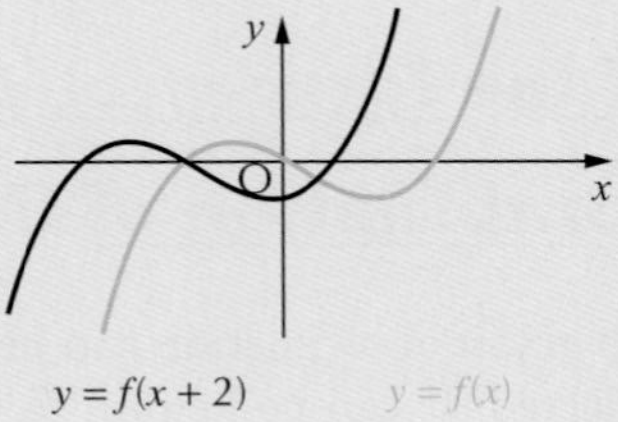 $y = f(x+2)$ $y = f(x)$

When a function is affected by more than one transformation the transformations must be applied in the correct order. The transformations follow the same 'order of operations' as normal numbers, i.e. in the same way that you multiply or divide before adding or subtracting, functions should be stretched, compressed or reflected before they are shifted vertically or horizontally.

Example 10.2 

The diagram shows the graph of a function $y = f(x)$.

Copy the diagram and on it sketch the graphs of:

(*a*) $y = f(x - 4)$; **2**

(*b*) $y = 2 + f(x - 4)$. **2**

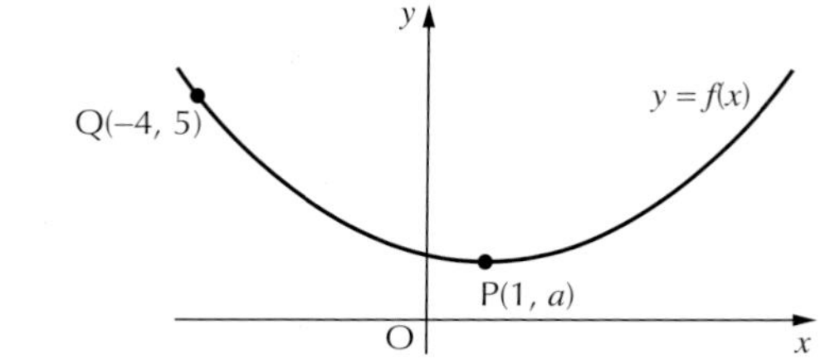

Hint

Rewriting functions like $y = 2 + f(x - 4)$ as $y = f(x - 4) + 2$ can make it easier to see that this is a shift to the right followed by a shift up.

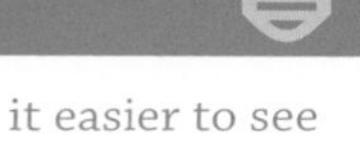

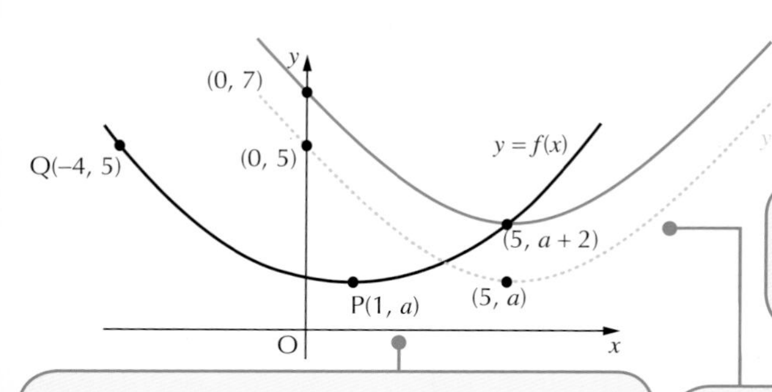

$y = 2 + f(x - 4)$

$y = f(x - 4)$

The −4 shifts the graph 4 units to the right. The 2 moves the graph 2 units up.

Marks for this question are awarded for:

- Moving the graph 4 units right (✓)
- Labelling all points on $y = f(x - 4)$ (✓)
- Moving the graph 2 units up (✓)
- Labelling all points on $y = 2 + f(x - 4)$ (✓)

The dotted line here shows $y = f(x - 4)$ and is the answer to part (a) of this question. Make sure you label all possible points on this graph. By completing part (a) all that remains to complete part (b) is to move the dotted curve 2 units vertically up.

Example 10.3 

Q The diagram below shows the graph of a quartic $y = h(x)$, with **stationary points** at $x = 0$ and $x = 2$.

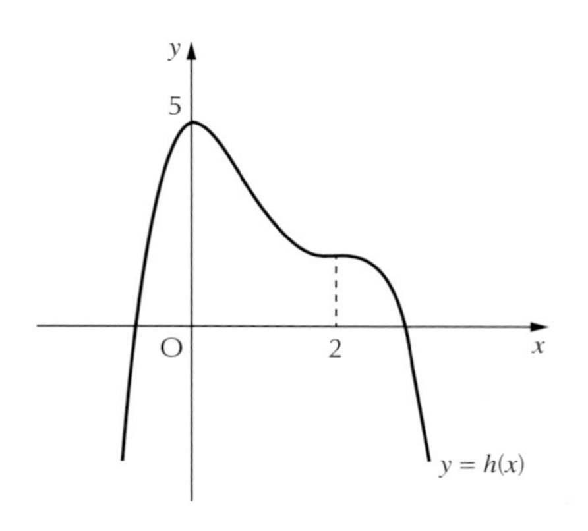

Hint

If you find yourself making several attempts at a graph, make sure it is clear which one is your final attempt; otherwise you may not get any marks.

Sketch the graph of $y = 2 - h(x)$. **3**

A $y = 2 - h(x)$

$y = -h(x) + 2$

Rewrite $y = 2 - h(x)$ as $y = -h(x) + 2$ to make it easier to see that this is a reflection in the x-axis followed by a shift up of 2 units.

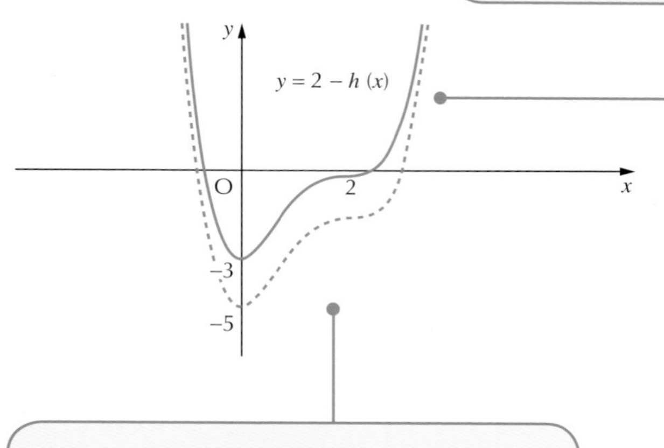

The dotted line shows $y = -h(x)$ and has been included here to let you see the first part of the translation, i.e. the graph has been reflected in the x-axis but has still to be shifted up. The dotted line is not required for full marks. However, to get full marks here you must include the x and y axes and label all points.

Marks for this question are awarded for:

- Reflecting the graph in the x-axis (✓)
- Moving the graph 2 units up (✓)
- Labelling all points on $y = 2 - h(x)$ (✓)

Sketching the graph of the derived function $y = f'(x)$ given the graph of $y = f(x)$

Sketching the derived function $y = f'(x)$ can be confusing but with plenty of practice it is possible for you to pick up some valuable marks here. A good starting point is to remember that stationary points on the original graph become **roots** on the derived graph.

Example 10.4 

The graph of the cubic function $y = f(x)$ is shown in the diagram. There are **turning points** at (1,1) and (3,5). Sketch the graph of $y = f'(x)$. **3**

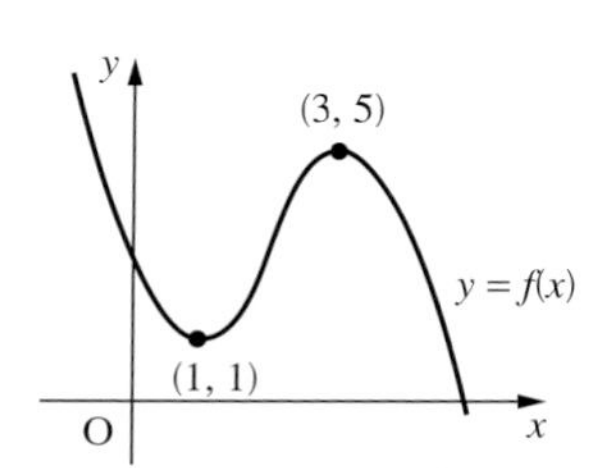

Hint

Stationary points of $f(x)$ become *roots* of $f'(x)$. A good first step here is to draw a set of axes exactly the same as those in the question, then mark the x-coordinates of any stationary points on your new x-axis as the roots. Next examine the **gradient** of $f(x)$, any *negative* sections will give sections on the graph of $f'(x)$ which are *below* the x-axis, any *positive* sections on $f(x)$ will give sections on $f'(x)$ which are *above* the x-axis.

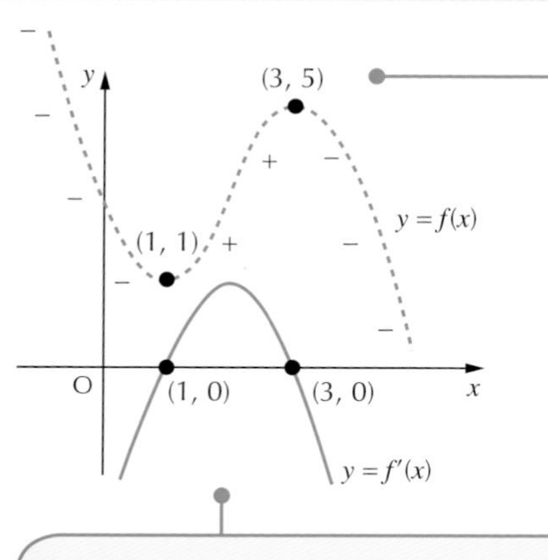

You don't have to include the original graph in your answer. It is shown here to allow you to see how the gradient of $f(x)$ corresponds to sections of $f'(x)$ that are above/below the x-axis and also that the roots of $f'(x)$ are directly below the turning points of $f(x)$. To get full marks you must include the x and y axes and you must label all points.

Marks for this question are awarded for:

- Correctly identifying the roots (✓)
- Drawing a maximum turning point between the roots (✓)
- Drawing a **parabola** which is symmetrical about the midpoint of the roots (✓)

Sketching the graph of an exponential or logarithmic function and finding the equation when given the graph

Providing $a > 0$ and $a \neq 1$, the graph of $y = \log_a x$ will pass through the points (0, 1) and (1, a). It will also approach but never cut the y-axis, making the y-axis a boundary line or *asymptote* of the graph. The graph of any function related to $f(x) = \log_a x$ can be drawn by thinking about the effects of the transformation on the points (1, 0) and (a, 1) and the boundary line or y-axis.

Example 10.5 

Q

The function f is of the form $f(x) = \log_b(x - a)$.
The graph of $y = f(x)$ is shown in the diagram.

(a) Write down the values of a and b. **2**

(b) State the domain of f. **1**

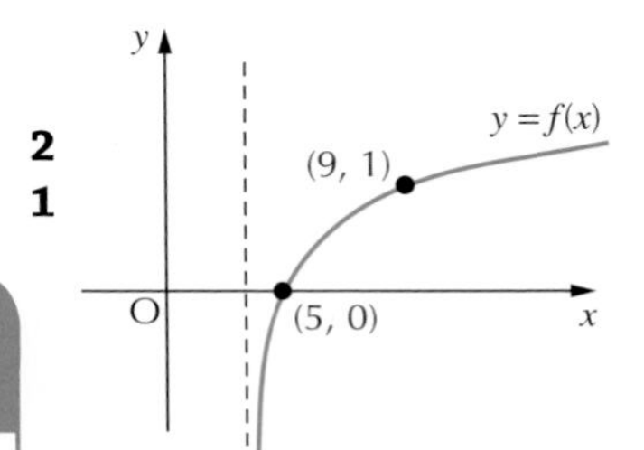

Hint

Unless they have been transformed in some way, all logarithmic graphs pass through the point (1, 0). Use this fact, along with the point where the given graph crosses the x-axis, to help you determine a (the horizontal shift). To find b you can substitute the value of a , along with the other given point into the equation.

A

(a)

$a = 4$ (✓) — Instead of passing through (1, 0) this graph passes through (5, 0). The graph has been shifted 4 units to the right meaning $a = 4$.

$\log_b (x - a) = y$

$\log_b (9 - 4) = 1$ — Substitute $x = 9$, $y = 1$ and $a = 4$ into the equation.

$\log_b 5 = 1$

$b = 5$ (✓) — Solve using the fact that $\log_5 5 = 1$.

(b)

Domain is $x > 4$ (✓)

Hint

In this question it is possible to answer part (b) without completing all of part (a), just don't panic and keep going! Remember, the doma**in** is all the x values that go **in**to the function.

The dotted line on the original diagram is called an asymptote. It is there to show you that, although our graph gets closer and closer to the dotted line, it never actually touches it. This tells us that x will always be greater than 4.

Sketch the graph or find the equation of a function of the form $y = a\sin(bx + d) + c$ or $y = a\cos(bx + d) + c$

When answering this type of question you will find that it helps to make a quick sketch of $y = \sin x$ or $y = \cos x$ and compare this with the graph in the question. The maximum and minimum values of the given graph will let you work out a (**amplitude**) and c (vertical shift). The number of complete waves between 0 and 360° (or 0 and 2π, if working with **radians**) will tell you b (the **frequency**). Finally, looking at where the graphs cross the x-axis will let you figure out d (phase angle).

Example 10.6 

The diagram shows the curve with equation of the form $y = \cos(x + a) + b$ for $0 \leqslant x \leqslant 2\pi$.

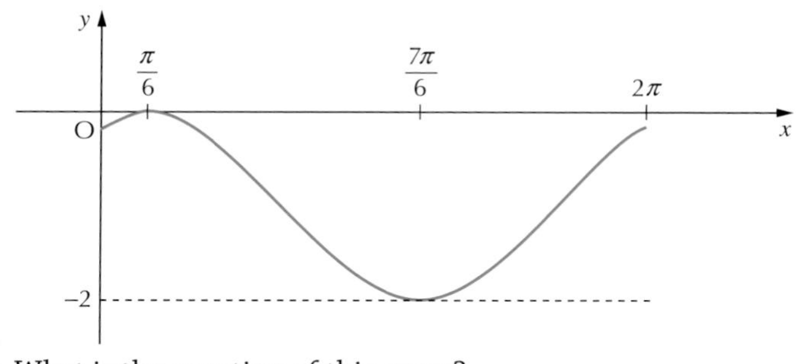

What is the equation of this curve? **2**

A $y = \cos(x + a) + b$

$a = -\frac{\pi}{6}$ (✓)

$b = -1$ (✓)

The equation of the curve is

$y = \cos\left(x - \frac{\pi}{6}\right) - 1.$

The graph of $y = \cos x$ usually has a minimum at −1. Since the given graph has a minimum at −2 (and there has been no change to the amplitude) then you know this graph has been shifted down by 1 unit.

Compare this graph to the graph of $y = \cos x$.

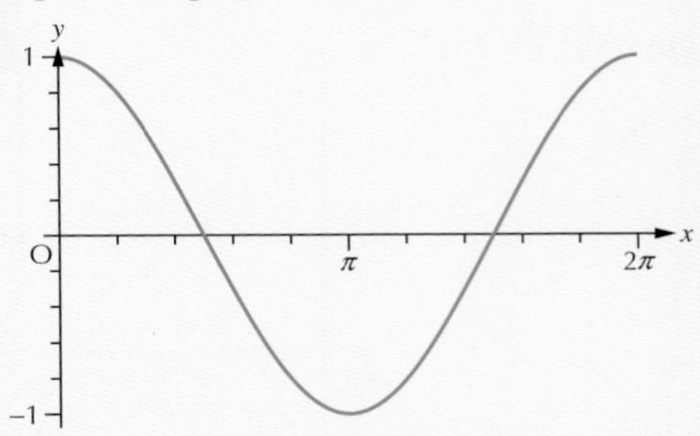

You can see it has been shifted down and it starts a little to the right of the y-axis.

Since the curve has been shifted to the right, the value of a will be negative.

Example 10.7 

Hint

In this example the equation $y = a \sin(bx) + c$ is used. a is the amplitude of the curve, b is the frequency (the number of waves between 0 and 2π), and c represents the vertical shift, i.e. how far up (or down) the y-axis the curve has been moved.

Q The diagram shows a sketch of part of the graph of a trigonometric function whose equation is of the form $y = a \sin(bx) + c$.
Determine the values of a, b and c. **3**

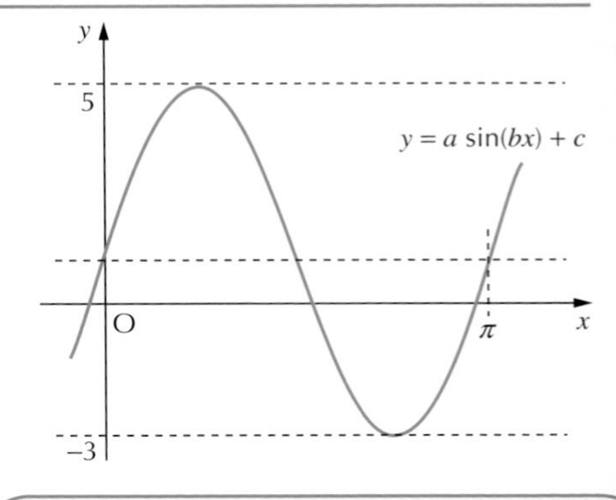

A $y = a \sin(bx) + c$

Amplitude $= \frac{\text{max} - \text{min}}{2} = \frac{5 - (-3)}{2} = 4$

$a = 4$ (✓)

$b = 2$ (✓)

$c = \text{max} - \text{amplitude}$

$c = 5 - 4$

$c = 1$ (✓)

Since the amplitude of this wave is 4, we might expect it to have a maximum at 4 and a minimum at −4. However, it has been shifted vertically and the maximum is actually at 5. The graph has been shifted up by 1 unit.

The total distance between the maximum (5) and the minimum (−3) is 5 − (−3) = 8. The amplitude (a) is half of this number.

Look carefully at the x-axis. One wave takes π radians to complete. This means that there will be two waves between 0 and 2π, therefore the frequency (b) is 2.

Find the maximum and minimum values of a trigonometric function of the form $y = a \sin(bx + d) + c$ or $y = a \cos(bx + d) + c$

Trigonometric functions can be used to describe anything that changes in a regular repeated way, like the depth of water in a harbour, the height of a big wheel at the fairground or the number of hours of daylight throughout a year. Your final answer might have some significance within the context of the question, although the way you approach these questions is usually the same.

Example 10.8 

The diagram shows an incomplete graph of $y = 5\sin\left(x - \frac{\pi}{4}\right) + 2$ for $0 \le x \le 2\pi$.

Find the coordinates of the maximum stationary point. **3**

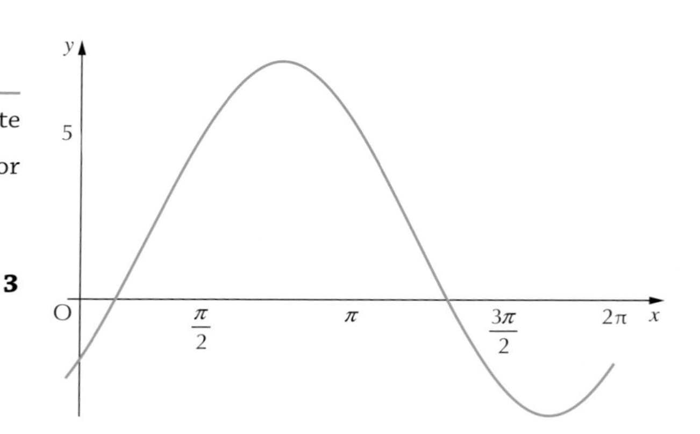

$$y = 5\sin\left(x - \frac{\pi}{4}\right) + 2$$

Maximum value is 7 (✓)

The 5 at the start of the equation tells us that this curve has an amplitude of 5. This would be the maximum value if it wasn't for the +2, which takes the maximum value of the function to 7.

$$x = \frac{\pi}{2} + \frac{\pi}{4} \quad (✓)$$

$$x = \frac{3\pi}{4}$$

Maximum stationary point at $\left(\frac{3\pi}{4}, 7\right)$ (✓)

You know from your knowledge of the graph of $y = \sin x$ that the first maximum value of $y = \sin x$ occurs when $x = \frac{\pi}{2}$.

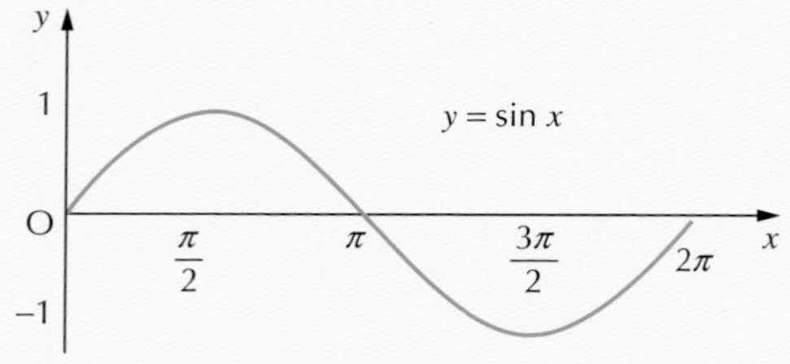

From the given equation, we know that every point on our graph has been shifted $\frac{\pi}{4}$ to the right, meaning that our maximum occurs at $x = \frac{\pi}{2} + \frac{\pi}{4}$.

Example 10.9 

Q For the function $f(x) = 2 - 5\sin\left(x - \frac{\pi}{3}\right)$ in the interval $0 \le x \le 2\pi$, determine which of the following statements are true **and justify your answer**. **3**

Statement A The maximum value of $f(x)$ is -3.

Statement B The maximum value of $f(x)$ is 7.

Statement C The maximum value occurs when $x = \frac{5\pi}{6}$.

Statement D The maximum value occurs when $x = \frac{11\pi}{6}$.

A

$y = 2 - 5\sin\left(x - \frac{\pi}{3}\right)$

$y = -5\sin\left(x - \frac{\pi}{3}\right) + 2$

Rewriting the function like this can make it easier to see the vertical shift of +2.

The amplitude of the curve is 5. The maximum value is $5 + 2 = 7$.

Statement B is correct (✓)

The −5 tells us that this curve has an amplitude of 5. This would be the maximum value if it wasn't for the +2, which takes the maximum value of the function to 7.

$x = \frac{3\pi}{2} + \frac{\pi}{3}$ (✓)

$x = \frac{9\pi}{6} + \frac{2\pi}{6}$

$x = \frac{11\pi}{6}$

Statement D is also correct. (✓)

Make sure you communicate your answer clearly.

You know from your knowledge of the graph of $y = \sin x$ that the first maximum value of $y = -5\sin x$ occurs when $x = \frac{3\pi}{2}$.

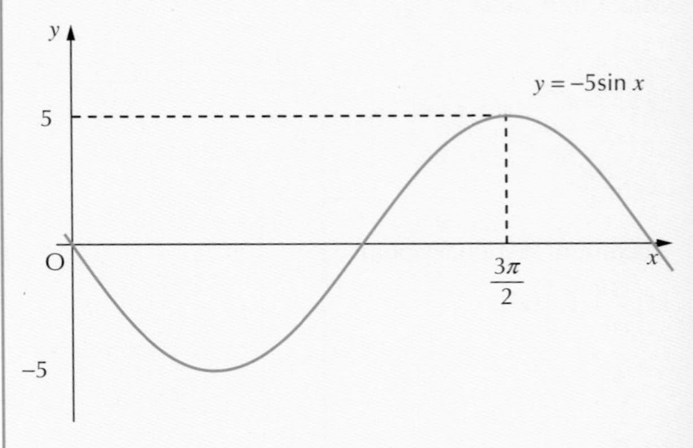

From the given equation $y = 2 - 5\sin\left(x - \frac{\pi}{3}\right)$, we know that every point on our graph has been shifted $\frac{\pi}{3}$ to the right, meaning that our maximum occurs at $x = \frac{3\pi}{2} + \frac{\pi}{3}$.

Find formulae for composite and inverse functions

The next example covers finding a composite function and finding an inverse function. The two processes will not always be together in the same question. Both rely on good algebra technique and finding the inverse function requires you to change the subject of a formula.

Example 10.10 

Functions f and g are defined on suitable domains by $f(x) = x^3 - 1$ and $g(x) = 3x + 1$.

(a) Find an **expression** for $k(x)$, where $k(x) = g(f(x))$. **2**

(b) If $h(k(x)) = x$, find an expression for $h(x)$. **3**

(a) $k(x) = g(f(x)) = g(x^3 - 1)$ (✓)

The entire f function becomes the input to the g function.

$k(x) = 3(x^3 - 1) + 1$ (✓)

Replace the x in $g(x) = 3x + 1$ with the entire f function, i.e. x is replaced with $x^3 - 1$.

(b) $h(x) = k^{-1}(x)$

Since $h(k(x)) = x$ you know that $h(x)$ must be the inverse of $k(x)$.

$k(x) = 3x^3 - 3 + 1$

$k(x) = 3x^3 - 2$

$y = 3x^3 - 2$

To keep the algebra simple, replace $k(x)$ with y then start to change the subject of the equation to x.

$3x^3 - 2 = y$

Rearrange the equation so that the x term is on the left-hand side.

$3x^3 = y + 2$ (✓)

$x^3 = \frac{y+2}{3}$

Adding 2 to both sides then dividing by 3 gets you to here.

$x = \sqrt[3]{\frac{y+2}{3}}$ (✓)

Taking the cube root of both sides completes the process.

$k^{-1}(x) = \sqrt[3]{\frac{x+2}{3}}$

Replace y with x and x with $k^{-1}(x)$.

$h(x) = \sqrt[3]{\frac{x+2}{3}}$ (✓)

Write out in full, remembering that we realised earlier that $h(x) = k^{-1}(x)$.

Hint

If you find a composite function that simplifies to x, then the two original functions will be the inverses of each other, i.e. one 'undoes' the other.

Complete the square where the coefficient of x^2 is non-unitary

Example 10.11 

Q $f(x)$ and $g(x)$ are functions, defined on the **set** of real numbers, such that

$f(x) = 1 - \frac{1}{2}x$ and $g(x) = 8x^2 - 3$.

(a) Given that $h(x) = g(f(x))$, show that $h(x) = 2x^2 - 8x + 5$. **3**

(b) Express $h(x)$ in the form $a(x+p)^2 + q$. **3**

(c) Hence, or otherwise, state the coordinates of the turning point on the graph of $y = h(x)$. **1**

(d) Sketch the graph of $y = h(x) + 3$, showing clearly the coordinates of the turning point and the y-axis intercept. **2**

A (a)

$h(x) = g(f(x))$

$h(x) = g\left(1 - \frac{1}{2}x\right)$

$h(x) = 8\left(1 - \frac{1}{2}x\right)^2 - 3$ (✓)

$h(x) = 8\left(1 - \frac{1}{2}x\right)\left(1 - \frac{1}{2}x\right) - 3$

$h(x) = 8\left(1 - \frac{1}{2}x - \frac{1}{2}x + \frac{1}{4}x^2\right) - 3$

$h(x) = 8\left(1 - x + \frac{1}{4}x^2\right) - 3$ (✓)

$h(x) = 8 - 8x + 2x^2 - 3$

$h(x) = 2x^2 - 8x + 5$ (✓)

Replace the x in $g(x) = 8x^2 - 3$ with the entire f function, i.e. x is replaced with $1 - \frac{1}{2}x$.

Take it one step at a time and write out the squaring in full.

Multiply the double brackets, taking care to realise that $\left(-\frac{1}{2}x\right) \times \left(-\frac{1}{2}x\right) = \frac{1}{4}x^2$.

Always simplify anything in brackets. In this case the x terms simplify (quite) nicely!

Multiply everything in the brackets by the 8 and the fractions disappear.

Combining the number terms and rearranging gives us what we were asked to show. Write your last line of working out exactly as it is shown in the question to make sure you get the final mark.

Hint

The good news about 'show that' questions is that you can get follow through marks even if you struggle with the 'show that' part of the question. This is because the result you need for the rest of the question is already given to you.

(b)

$h(x) = 2x^2 - 8x + 5$

Remember, you can complete part (b) without having finished part (a), just make sure you use the result given in the question.

$h(x) = 2(x^2 - 4x) + 5$ (✓)

Identify and remove a common **factor** from the first two terms.

$h(x) = 2[(x - 2)^2 - 4] + 5$ (✓)

Complete the square. See Chapter 6 (Essential skills) for a recap on how to complete the square.

$h(x) = 2(x - 2)^2 - 8 + 5$

Multiply everything *inside* the square bracket by the 2, note that the 5 is not multiplied.

$h(x) = 2(x - 2)^2 - 3$ (✓)

Simplify the number terms.

(c)

$h(x) = 2(x - 2)^2 - 3$
Turning point at $(2, -3)$ (✓)

Read off the turning point from your answer to part (b).

Hint

The word "hence" tells you that you should be able to use your result from a previous part of the question to answer this part.

(d)

$h(x) = 2x^2 - 8x + 5$
$h(x) + 3 = 2x^2 - 8x + 8$

Hint

Use the original equation to determine the y – intercept. When written in the form $y = ax^2 + bx + c$, the y – intercept will always be at c (since $x = 0$ at the intercept).

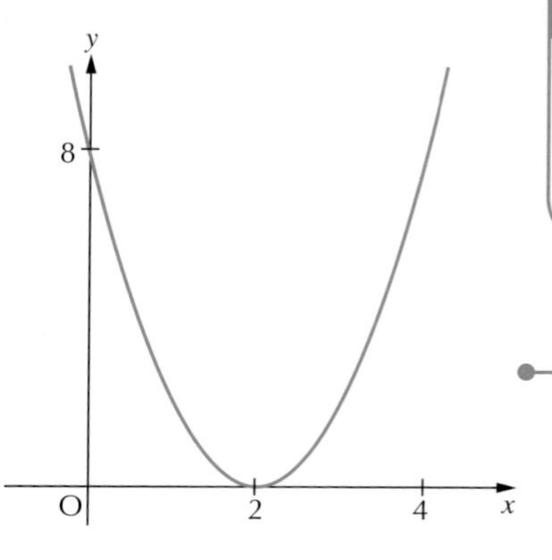

$h(x)$ has a y – intercept at 5 and a turning point at $(2, -3)$. $h(x) + 3$ is shifted 3 units up the y-axis and will have a y – intercept at 8 and a turning point at $(2, 0)$. You know the graph will be a ∪-shaped parabola because the coefficient of x^2 in the equation is greater than zero.

Marks for this question are awarded for:

- Drawing a parabola with an appropriately labelled minimum turning point (✓)
- Correctly marking on the y – intercept (✓)

Vectors

Before you start this chapter you should be able to:

- Determine the coordinates of a point from a diagram in 3D (Chapter 6)
- Add/subtract 2D **vectors** using **directed line segments** (Chapter 6)
- Add/subtract 3D vectors using **components** (Chapter 6)
- Find the **magnitude** of a vector (Chapter 6).

This chapter covers:

- Working with **parallel** vectors and **collinearity**
- Finding the coordinates of a point which divides a line in a given **ratio**
- Evaluating and applying properties of the **scalar product**
- Working with the resultant of vector pathways in 3D
- Problems involving the angle between two vectors
- Working with **perpendicular vectors**
- Working with vectors defined in terms of the **unit vectors i, j and k.**

Working with parallel vectors and collinearity

Gradients can be used to show that lines are parallel in two dimensions. This method will not work in three dimensions, where the concept of gradient doesn't exist. You can show that two vectors are parallel by demonstrating that the components of one vector are a **scalar** multiple of the components of the other, i.e. if $\mathbf{a} = k\mathbf{b}$ for some scalar $k \neq 0$ then $\mathbf{a}$ is parallel to $\mathbf{b}$ and vice-versa.

Points are said to be collinear if they lie on the same straight line. Remember, only points can be collinear, not vectors. See Chapter 6 (Essential skills) for more on working with vectors.

Example 11.1 

(a) Show that the points A(−7, −8, 1), T(3, 2, 5) and B(18, 17, 11) are collinear. **3**

(b) Find the ratio in which T divides AB. **1**

(a) $\overrightarrow{AT} = \mathbf{t} - \mathbf{a} = \begin{pmatrix} 3 \\ 2 \\ 5 \end{pmatrix} - \begin{pmatrix} -7 \\ -8 \\ 1 \end{pmatrix} = \begin{pmatrix} 10 \\ 10 \\ 4 \end{pmatrix} = 2\begin{pmatrix} 5 \\ 5 \\ 2 \end{pmatrix}$ (✓)

> It is a good idea to take out the highest common **factor**. This will let you see that one vector is a multiple of the other.

$\overrightarrow{TB} = \mathbf{b} - \mathbf{t} = \begin{pmatrix} 18 \\ 17 \\ 11 \end{pmatrix} - \begin{pmatrix} 3 \\ 2 \\ 5 \end{pmatrix} = \begin{pmatrix} 15 \\ 15 \\ 6 \end{pmatrix} = 3\begin{pmatrix} 5 \\ 5 \\ 2 \end{pmatrix}$

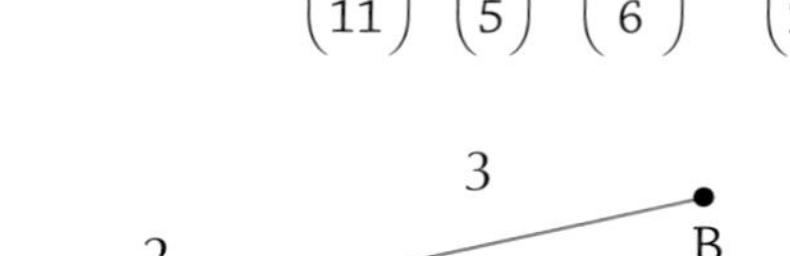

Hint

In questions like this, if you are going to find the ratio in which T divides AB you should make T your common point.

$\overrightarrow{AT} = \frac{2}{3}\overrightarrow{TB}$ (✓)

> You can tell from the components that $\overrightarrow{AT}$ is smaller than $\overrightarrow{TB}$. By looking at the factors, the components, or by making a sketch, you can see that $\overrightarrow{AT} = \frac{2}{3}\overrightarrow{TB}$.

$\overrightarrow{AT}$ and $\overrightarrow{TB}$ are parallel and since there is a common point A, B and T are collinear. (✓)

> Communication is very important here. To get the mark you *must* use the words 'parallel', 'common point' *and* 'collinear'.

(b) T divides AB in the ratio 2 : 3. (✓)

> It is easy to get confused here. Just remember that, since AT is shorter than TB, then T must be closer to A. That is why the '2' comes first in the ratio. If you used T as your common point then the ratio comes from the common factors.

Finding the coordinates of a point which divides a line in a given ratio

The best advice for this type of question is to make a sketch. This will help you to select the correct ratio. In the following example we have chosen to use an algebraic approach when working with the vectors. This method follows algebraic steps you should be familiar with and means that any numbers are only substituted near the end of the working. You can also use this method in circles questions when you are trying to determine the coordinates of the **centre** or to find a point on the circle.

Example 11.2

Q The point Q divides the line joining P(–1, –1, 0) to R(5, 2, –3) in the ratio 2 : 1.

Find the coordinates of Q. **3**

A

P —— 2 —— Q —— 1 —— R

A quick sketch can be really helpful. It lets you see that the journey from P to R can be thought of as three 'parts'.

$\overrightarrow{PQ} = \frac{2}{3}\overrightarrow{PR}$ (✓)

Although we don't know Q, we do know P and R and can set up this equation since Q is 'two-thirds' of the way from P to R.

$3\overrightarrow{PQ} = 2\overrightarrow{PR}$

$3(\mathbf{q} - \mathbf{p}) = 2(\mathbf{r} - \mathbf{p})$

Multiply both sides by 3 to remove the fraction, then write in terms of **position vectors**, e.g. **q**, **p** and **r**.

$3\mathbf{q} - 3\mathbf{p} = 2\mathbf{r} - 2\mathbf{p}$

$3\mathbf{q} = 2\mathbf{r} + \mathbf{p}$

Remove the brackets by multiplying, then rearrange so that the term we are trying to find, in this case **q**, is on the left-hand side.

$3\mathbf{q} = 2\begin{pmatrix} 5 \\ 2 \\ -3 \end{pmatrix} + \begin{pmatrix} -1 \\ -1 \\ 0 \end{pmatrix}$ (✓)

Replace **r** and **p** with the position vectors given in the question.

$3\mathbf{q} = \begin{pmatrix} 10 \\ 4 \\ -6 \end{pmatrix} + \begin{pmatrix} -1 \\ -1 \\ 0 \end{pmatrix} = \begin{pmatrix} 9 \\ 3 \\ -6 \end{pmatrix}$

$\mathbf{q} = \begin{pmatrix} 3 \\ 1 \\ -2 \end{pmatrix}$

Simplify to give 3q, then divide by 3 to get the position vector of **q**.

Q (3,1, –2) (✓)

Remember to write Q as a coordinate to get the final mark.

Evaluating and applying properties of the scalar product

The scalar product involves vectors but, as its name suggests, the result is not a vector but a scalar (a number). The scalar product is found by multiplying the magnitudes of two vectors and the cosine of the angle between them. The scalar product is defined as $\mathbf{a}.\mathbf{b} = |\mathbf{a}||\mathbf{b}|\cos\theta$ where θ is the angle between the vectors when placed 'tail-to-tail' or 'nose-to-nose'. Sometimes it is necessary to redraw the vectors so they are positioned correctly.

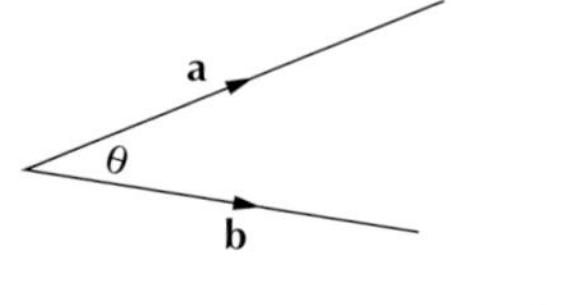

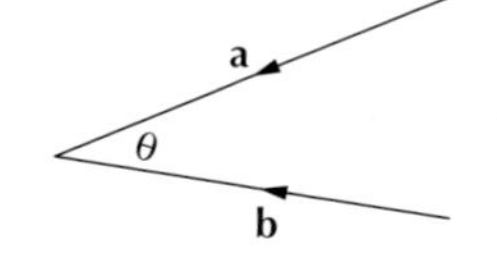

When you are told that two vectors are 'perpendicular' this also tells you that the scalar product of the two vectors will be equal to zero.

Example 11.3

An equilateral triangle with sides of length 3 units is shown.

Vector $\mathbf{r}$ is 2 units long and is perpendicular to both vectors $\mathbf{p}$ and $\mathbf{q}$.

Calculate the value of the scalar product $\mathbf{p}.(\mathbf{p}+\mathbf{q}+\mathbf{r})$ **4**

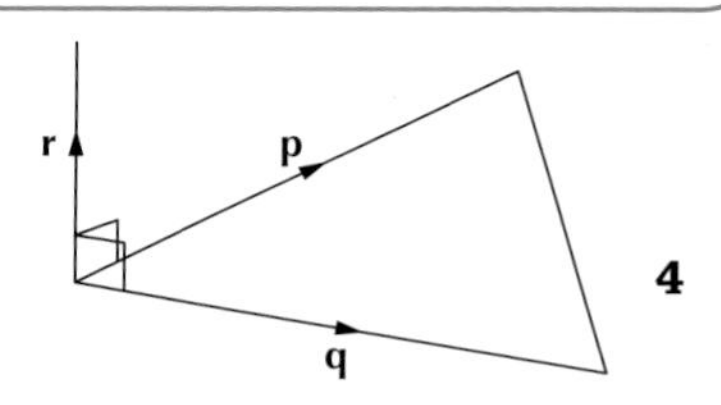

$\mathbf{p}.(\mathbf{p}+\mathbf{q}+\mathbf{r})$

$= \mathbf{p}.\mathbf{p} + \mathbf{p}.\mathbf{q} + \mathbf{p}.\mathbf{r}$ (✓)

Remove the brackets. The scalar product follows the 'distributive law' like regular multiplication.

$= |\mathbf{p}||\mathbf{p}|\cos\theta_1 + |\mathbf{p}||\mathbf{q}|\cos\theta_2 + |\mathbf{p}||\mathbf{r}|\cos\theta_3$

Use your formula sheet to write out these three expansions of the scalar product. θ_1 is the angle between $\mathbf{p}$ and $\mathbf{p}$ (which is zero), θ_2 is the angle between $\mathbf{p}$ and $\mathbf{q}$ and θ_3 is the angle between $\mathbf{p}$ and $\mathbf{r}$.

$= |\mathbf{p}||\mathbf{p}|\cos 0 + |\mathbf{p}||\mathbf{q}|\cos 60 + |\mathbf{p}||\mathbf{r}|\cos 90$

The angle between $\mathbf{p}$ and $\mathbf{q}$ is 60° (since the triangle is equilateral) and θ_3, the angle between $\mathbf{p}$ and $\mathbf{r}$, is 90°, as shown in the diagram.

$= |\mathbf{p}||\mathbf{p}|(1) + |\mathbf{p}||\mathbf{q}|\left(\frac{1}{2}\right) + |\mathbf{p}||\mathbf{r}|(0)$

Use your knowledge of the graph of $y = \cos x$ and the **exact value** triangles to replace cos 0, cos 60 and cos 90. See Chapter 9 (Manipulating trigonometric **expressions**) for more on exact values.

$= (3)(3)(1) + (3)(3)\left(\frac{1}{2}\right) + (3)(2)(0)$

Substitute the magnitudes of $\mathbf{p}$, $\mathbf{q}$ and $\mathbf{r}$ given in the question.

$= 9 + \frac{9}{2}$ (✓)(✓)

Work out each term. Remember, anything multiplied by zero is zero!

$= \frac{27}{2}$ (✓)

Write your answer as a single fraction.

Working with the resultant of vector pathways in 3D/ Problems involving the angle between two vectors

The vector pathways in the diagram show three different routes from A to C:

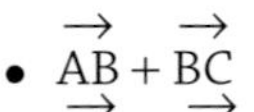

- $\overrightarrow{AB} + \overrightarrow{BC}$
- $\overrightarrow{AD} + \overrightarrow{DC}$
- $\overrightarrow{AC}$

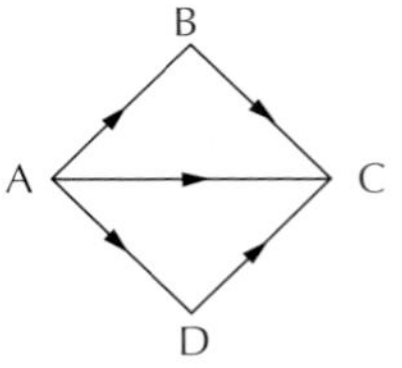

In vector terms each is equivalent to the other and to the vector $\overrightarrow{AC}$, which is the shortest path from A to C. $\overrightarrow{AC}$ is the resultant of each of the other two vector pathways. All unbroken pathways made from directed line segments which start at A and end at C are equivalent to $\overrightarrow{AC}$.

Example 11.4

Q

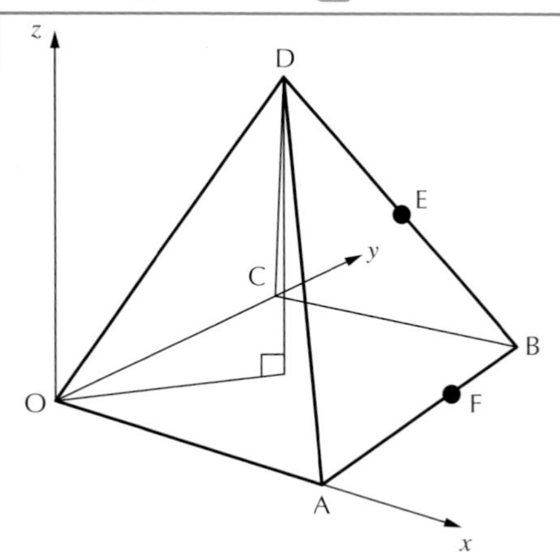

Hint

A pyramid where the apex (top) lies directly above the centre of the base is called a right pyramid.

A square based right pyramid is shown in the diagram.

Square OABC has a side length of 60 units with edges OA and OC lying on the x-axis and y-axis respectively.

The coordinates of D are (30, 30, 80).

E is the midpoint of BD and F divides AB in the ratio 2 : 1.

(a) Find the coordinates of E and F. **2**

(b) Calculate $\overrightarrow{ED}.\overrightarrow{EF}$. **2**

(c) Hence, or otherwise, calculate the size of angle DEF. **4**

(a) D(30, 30, 80) and B (60, 60, 0)

From the diagram we can see that to get from the origin to point B we go 60 units along the x-axis, 60 along the y-axis and zero up the z-axis. Hence B in coordinate form is (60, 60, 0).

$$\overrightarrow{BD} = \mathbf{d} - \mathbf{b} = \begin{pmatrix} 30 \\ 30 \\ 80 \end{pmatrix} - \begin{pmatrix} 60 \\ 60 \\ 0 \end{pmatrix} = \begin{pmatrix} -30 \\ -30 \\ 80 \end{pmatrix}$$

To find E, which is halfway along BD, we first need to find BD.

$$\mathbf{e} = \mathbf{b} + \frac{1}{2}\overrightarrow{BD}$$

If we add half of the 'journey' BD to the position vector **b**, we will get the position vector **e**.

$$= \begin{pmatrix} 60 \\ 60 \\ 0 \end{pmatrix} + \frac{1}{2}\begin{pmatrix} -30 \\ -30 \\ 80 \end{pmatrix} = \begin{pmatrix} 45 \\ 45 \\ 40 \end{pmatrix}$$

This is the position vector **e**, i.e. the journey from the origin to the point E.

E(45, 45, 40) (✓)

Write E in coordinate form.

A(60, 0, 0) and B(60, 60, 0)

From the diagram we can see that to get from the origin to point A we go 60 units along the x-axis, zero along the y-axis and zero up the z-axis. Hence A in coordinate form is (60, 0, 0).

$$\overrightarrow{AB} = \mathbf{b} - \mathbf{a} = \begin{pmatrix} 60 \\ 60 \\ 0 \end{pmatrix} - \begin{pmatrix} 60 \\ 0 \\ 0 \end{pmatrix} = \begin{pmatrix} 0 \\ 60 \\ 0 \end{pmatrix}$$

To find F, which is two-thirds along AB, we first need to find AB.

$$\mathbf{f} = \mathbf{a} + \frac{2}{3}\overrightarrow{AB}$$

If we add two-thirds of the 'journey' AB to the position vector **a**, we will get the position vector **f**.

$$= \begin{pmatrix} 60 \\ 0 \\ 0 \end{pmatrix} + \frac{2}{3}\begin{pmatrix} 0 \\ 60 \\ 0 \end{pmatrix} = \begin{pmatrix} 60 \\ 40 \\ 0 \end{pmatrix}$$

This is the position vector **f**, i.e. the journey from the origin to the point F.

F(60, 40, 0) (✓)

Write F in coordinate form.

(b)

$$\overrightarrow{ED} = \mathbf{d} - \mathbf{e} = \begin{pmatrix} 30 \\ 30 \\ 80 \end{pmatrix} - \begin{pmatrix} 45 \\ 45 \\ 40 \end{pmatrix} = \begin{pmatrix} -15 \\ -15 \\ 40 \end{pmatrix}$$

$$\overrightarrow{EF} = \mathbf{f} - \mathbf{e} = \begin{pmatrix} 60 \\ 40 \\ 0 \end{pmatrix} - \begin{pmatrix} 45 \\ 45 \\ 40 \end{pmatrix} = \begin{pmatrix} 15 \\ -5 \\ -40 \end{pmatrix} \quad (\checkmark)$$

Use your answers to part (a) to find $\overrightarrow{ED}$ and $\overrightarrow{EF}$.

$$\overrightarrow{ED}.\overrightarrow{EF} = (-15)(15) + (-15)(-5) + (40)(-40) = -1750 \quad (\checkmark)$$

Evaluate the scalar product using the formula, $\mathsf{a.b} = a_1b_1 + a_2b_2 + a_3b_3$, which is given to you in the formula sheet.

(c)

$$\cos DEF = \frac{\overrightarrow{ED}.\overrightarrow{EF}}{\left|\overrightarrow{ED}\right|\left|\overrightarrow{EF}\right|} \quad (\checkmark)$$

To calculate the size of angle DEF rearrange the formula $\mathsf{a.b} = |\mathsf{a}||\mathsf{b}|\cos\theta$ given to you in your formula sheet but, to get all the marks, make sure you use the letters D, E and F.

$$\left|\overrightarrow{ED}\right| = \sqrt{(-15)^2 + (-15)^2 + 40^2} = \sqrt{2050} \quad (\checkmark)$$

$$\left|\overrightarrow{EF}\right| = \sqrt{(15)^2 + (-5)^2 + (-40)^2} = \sqrt{1850} \quad (\checkmark)$$

Calculate the magnitudes of $\overrightarrow{ED}$ and $\overrightarrow{EF}$. You don't have to simplify the surds since the values are going to be used in the next stages of the calculation. However, don't evaluate the surds, e.g. $\sqrt{2050} = 45{\cdot}276...$, as you may lose marks due to rounding errors.

$$\cos DEF = \frac{-1750}{\sqrt{2050}\sqrt{1850}}$$

$$DEF = \cos^{-1}\left(\frac{-1750}{\sqrt{2050}\sqrt{1850}}\right)$$

$$DEF = 153.977...$$

$$= 154° \quad (\checkmark)$$

Substitute the values you have calculated for the scalar product and the magnitudes into the formula.

If this was a normal trigonometric equation, with answers between 0 and 360°, then you would find a first quadrant angle, then use the four quadrant diagram to determine where $y = \cos x$ is negative and so on. However, in this situation there will only be one correct angle and keying the expression into your calculator will give it straight away.

Working with perpendicular vectors/Working with vectors defined in terms of the unit vectors i, j and k

i, **j** and **k** are unit vectors, i.e. they have a magnitude of one. They lie on the x, y and z axes respectively and are perpendicular to each other.

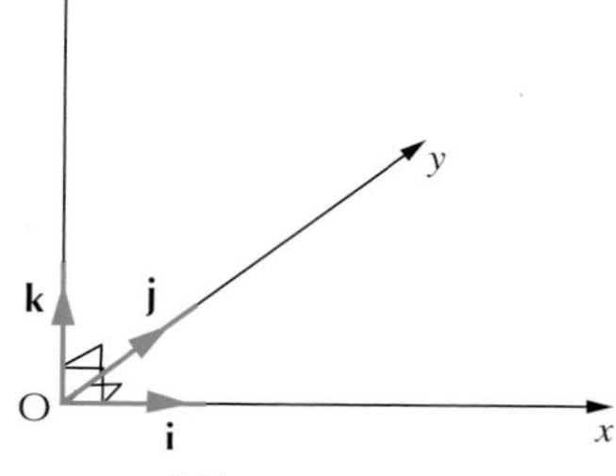

$$\mathbf{i} = \begin{pmatrix} 1 \\ 0 \\ 0 \end{pmatrix}, \ \mathbf{j} = \begin{pmatrix} 0 \\ 1 \\ 0 \end{pmatrix} \text{ and } \mathbf{k} = \begin{pmatrix} 0 \\ 0 \\ 1 \end{pmatrix}$$

Every vector can be expressed in terms of the unit vectors as follows:

$$\mathbf{u} = \begin{pmatrix} 2 \\ 1 \\ -3 \end{pmatrix}$$

$$\mathbf{u} = 2\begin{pmatrix} 1 \\ 0 \\ 0 \end{pmatrix} + 1\begin{pmatrix} 0 \\ 1 \\ 0 \end{pmatrix} + (-3)\begin{pmatrix} 0 \\ 0 \\ 1 \end{pmatrix}$$

$$\mathbf{u} = 2\mathbf{i} + \mathbf{j} - 3\mathbf{k}$$

Hint

If two non-zero vectors have a scalar product of zero then the vectors are perpendicular to each other, i.e. the angle between the vectors is 90°. Since cos 90° = 0 then $\mathbf{a}.\mathbf{b} = |\mathbf{a}||\mathbf{b}|\cos 90° = 0$.

Example 11.5

Q Vectors **u** and **v** are defined by $\mathbf{u} = 3\mathbf{i} + 2\mathbf{j}$ and $\mathbf{v} = 2\mathbf{i} - 3\mathbf{j} + 4\mathbf{k}$.

Determine whether or not **u** and **v** are perpendicular to each other. **2**

A $\mathbf{u} = 3\mathbf{i} + 2\mathbf{j}$ and $\mathbf{v} = 2\mathbf{i} - 3\mathbf{j} + 4\mathbf{k}$ — If **u** and **v** are perpendicular then $\mathbf{u}.\mathbf{v} = 0$.

$\mathbf{u}.\mathbf{v} = (3)(2) + (2)(-3) + (0)(4)$ — Substitute the **coefficients** of **i**, **j** and **k** into the scalar product formula given on your formula sheet. Note that the coefficient of **k** in vector **u** is zero.

$\mathbf{u}.\mathbf{v} = 6 + (-6) + 0 = 0$ (✓) — Carefully evaluate the scalar product. Remember, anything multiplied by zero is zero!

$\mathbf{u}.\mathbf{v} = 0$ so the vectors **u** and **v** are perpendicular. (✓) — Communicate your answer fully.

Unit 2: Relationships and calculus

Polynomials

Before you start this chapter you should be able to:

- **Factorise** an algebraic **expression** using the highest common **factor** (Chapter 6)
- Evaluate $f(x)$ at $x = a$ (Chapter 6)
- Complete the square (Chapter 10)
- Solve quadratic equations by factorising and using the quadratic formula (Chapter 6)
- Use the **discriminant** to determine the nature of the **roots** of a quadratic equation.

This chapter covers:

- Factorising and solving a cubic polynomial with a unitary x^3 **coefficient**, e.g. factorise and solve $x^3 - 2x^2 - x + 2 = 0$
- Factorising and solving a cubic or quartic with non-unitary coefficient of the highest power, e.g. factorise and solve $2x^4 + x^3 - 6x^2 + x + 2 = 0$
- Determining the equation of a polynomial from its graph
- Finding the point(s) of **intersection** of a straight line and a curve or of two curves
- Using the discriminant to determine an unknown coefficient in a quadratic equation given the nature of the roots.

Finding the point(s) of intersection of a straight line and a curve or of two curves

Example 12.1 

Q Find the coordinates of the points of intersection of the curve $y = x^3 - 2x^2 + x + 4$ and the line $y = 4x + 4$. **5**

A $y = x^3 - 2x^2 + x + 4$ and $y = 4x + 4$

Hint

To find where two lines or curves intersect, make sure both equations have y as the subject then set them equal to each other.

These equations represent a cubic curve and a straight line as shown:

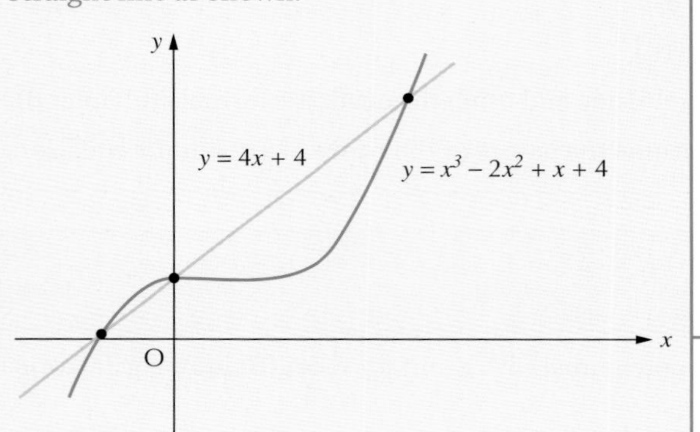

First we want to find the **values** of x at the three points of intersection.

$x^3 - 2x^2 + x + 4 = 4x + 4$ (✓) — Set the equations equal to each other.

$x^3 - 2x^2 - 3x = 0$ (✓) — Rearrange and make equal to zero. The '= 0' must appear to ensure full marks.

$x(x^2 - 2x - 3) = 0$ — Start to factorise, remember to always look for a common factor first.

$x(x - 3)(x + 1) = 0$ (✓) — Complete the factorisation, remember to check that your factors multiply back to give the original expression.

$x = 0, x = -1, x = 3$ (✓) — Solve, these are the x-coordinates of the points where the line and the curve intersect.

When $x = 0$, $y = 4(0) + 4 = 4$

When $x = -1$, $y = 4(-1) + 4 = 0$

When $x = 3$, $y = 4(3) + 4 = 16$

Substitute each x value back into either of the original equations to find the corresponding y values. It makes sense to substitute into the straight line equation as it is simpler.

Points of intersection at (0,4), (–1,0) and (3,16) (✓)

Summarise your answer and make sure you give coordinates!

Factorising a cubic polynomial with a unitary x^3 coefficient

There are a few alternative methods for factorising polynomials. This chapter will use the synthetic division method, which is the most common. Always use the method you are most comfortable with.

Example 12.2

(a) Show that $(x + 1)$ is a factor of $x^3 - 21x - 20$. **3**

(b) Factorise $x^3 - 21x - 20$ fully. **2**

(a)

	x^3	x^2	x	c	
–1	1	0	–21	–20	(✓)
		–1	1	20	
	1	–1	–20	0 (boxed)	(✓)

Decreasing powers of x. Here c represents the **constant** (number) term.

Coefficients of x^3, x^2, x and the constant. If a term is missing you **must** include a zero, in this case the x^2 is missing.

Add vertically then multiply diagonally from left to right.

The number in the box is the remainder when the polynomial is divided by $(x + 1)$. A zero here means $(x + 1)$ is a factor.

$(x + 1)$ is a factor as the remainder is zero. (✓)

Communication is essential here to get the mark.

(b)

$x^3 - 21x - 20$

$= (x + 1)(x^2 - x - 20)$ (✓)

Use the numbers from the bottom row of the synthetic division table as the coefficients of the quadratic factor, $x^2 - x - 20$.

$= (x + 1)(x + 4)(x - 5)$ (✓)

Factorise the quadratic factor to leave the **function** in fully factorised form.

Hint

There are some situations when you might use synthetic division and obtain a quadratic factor which you can't factorise. This doesn't necessarily mean you have made a mistake, it just means that there are no more solutions to the polynomial. If this happens you must show that $b^2 - 4ac < 0$ and explain that there are no further solutions.

Determining the equation of a polynomial from its graph

The previous examples have involved factorising a polynomial to find the roots (the points of intersection with the x-axis). This process can be reversed, allowing us to determine the equation of a polynomial using the information given in the graph. The values of the roots can be used to express the polynomial in factorised form, i.e. $y = k(x - a)(x - b)(x - c)$ where k is a numerical common factor ($k \neq 0$) which affects the **amplitude** of the graph. Note that any or all of a, b and c could be zero.

Example 12.3 

Q The diagram shows the curve with equation $y = f(x)$, where $f(x) = kx(x + a)(x + b)$.
The curve passes through (–1, 0), (0, 0), (1, 2) and (2, 0).

Find the values of a, b and k. **3**

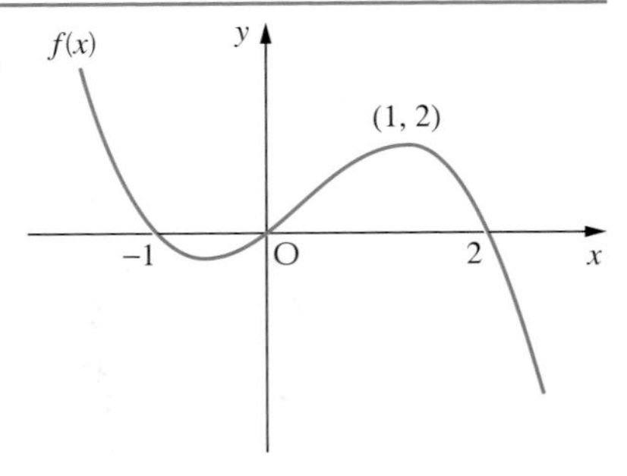

A

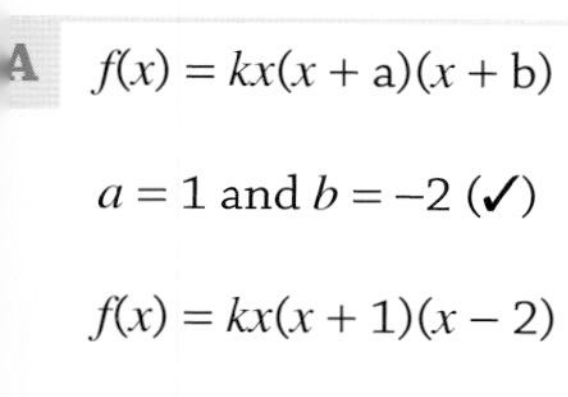

$f(x) = kx(x + a)(x + b)$

$a = 1$ and $b = -2$ (✓)

$f(x) = kx(x + 1)(x - 2)$

$k(1)(1 + 1)(1 - 2) = 2$ (✓)

$k(2)(-1) = 2$

$-2k = 2$

$k = -1$ (✓)

Use the roots indicated on the diagram to find the factors. Be careful with the signs. Here, since $x = -1$ and $x = 2$ are solutions (or roots) then $(x + 1)$ and $(x - 2)$ are factors.

Substitute your values for a and b into the function.

Since (1,2) is a point on the line, when x is replaced with 1, the function has the value 2.

Solve to find k.

Factorising a polynomial with non-unitary x^3 coefficient

You may be asked to factorise a polynomial with a 'non-unitary' x^3 coefficient, i.e. there will be a number (other than one) in front of the x^3 term. Polynomials of this type are factorised (and solved) in the same way as those with unitary x^3 coefficients.

Example 12.4 

Q

For the polynomial $6x^3 + 7x^2 + ax + b$,

- $x + 1$ is a factor
- 72 is the remainder when it is divided by $x - 2$.

(a) Determine the values of a and b. **4**

(b) Hence factorise the polynomial completely. **3**

Hint

When you are asked to find two unknowns, always consider simultaneous equations as a possible method. In this question you have to create two equations using the facts before solving them to find a and b.

A (a)

Since $(x + 1)$ is a factor, $f(-1) = 0$

$f(-1) = 6(-1)^3 + 7(-1)^2 + a(-1) + b = 0$ (✓)

Since $(x + 1)$ is a factor then substituting the root $x = -1$ into the function gives a value of zero.

$-6 + 7 - a + b = 0$

$-a + b = -1$

Simplify to get an equation in a and b.

$f(2) = 72$

$6(2)^3 + 7(2)^2 + a(2) + b = 72$ (✓)

Since dividing by $(x - 2)$ gives a remainder of 72 then substituting $x = 2$ into the function gives a value of 72.

$6(8) + 7(4) + 2a + b = 72$

$2a + b = -4$

Simplify to get a second equation in a and b.

$-a + b = -1$

$3a = -3$

$a = -1$ (✓)

Solve the equations simultaneously to find a.

$b = a - 1 = -1 - 1 = -2$ (✓)

Substitute the answer for a into the first equation to find b.

(b)

Keyword

The word 'hence' in the question tells you to use your answer from part (a) when answering part (b).

$6x^3 + 7x^2 - x - 2$

Write out the polynomial including the values for a and b found in part (a).

	x^3	x^2	x	c
-1	6	7	-1	-2 (✓)
		-6	-1	2
	6	1	-2	0

We were told in the question that $(x + 1)$ was a factor so we know $x = -1$ is a root. Factorise using synthetic division or any method you are familiar with.

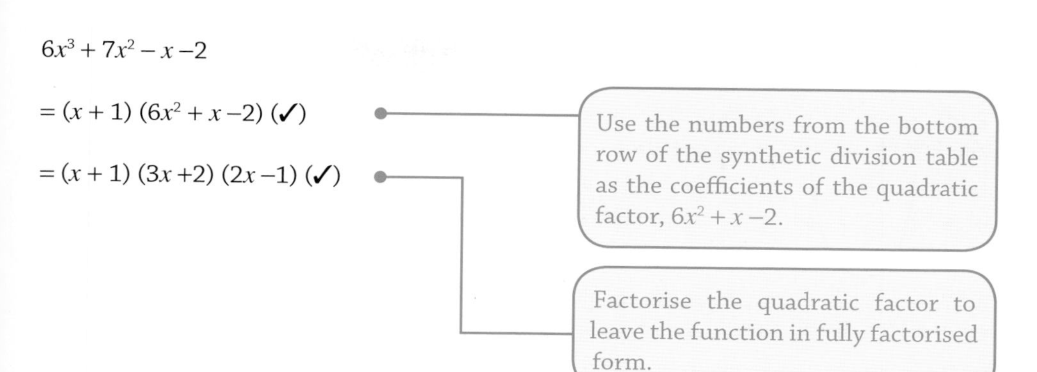

$6x^3 + 7x^2 - x - 2$

$= (x + 1)(6x^2 + x - 2)$ (✓)

$= (x + 1)(3x + 2)(2x - 1)$ (✓)

Using the discriminant to determine an unknown coefficient in a quadratic given the nature of the roots

Questions which talk about 'roots' or the 'nature of roots' will most likely require the discriminant, i.e. $b^2 - 4ac$. Remember there are three possibilities for a quadratic equation:

- Distinct real roots, i.e. $b^2 - 4ac < 0$

$b^2 - 4ac > 0$

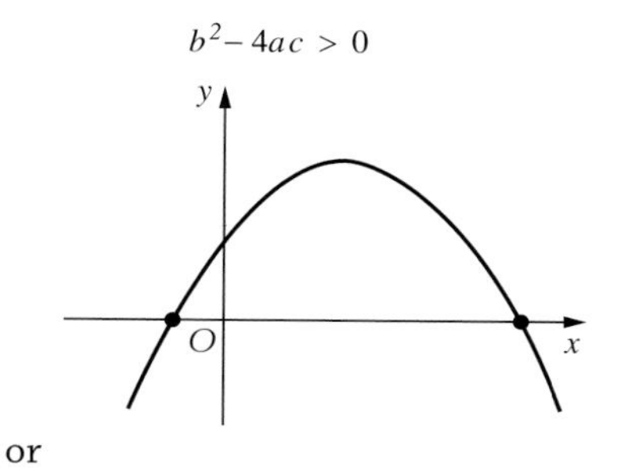

or

- Equal real roots, i.e. $b^2 - 4ac = 0$

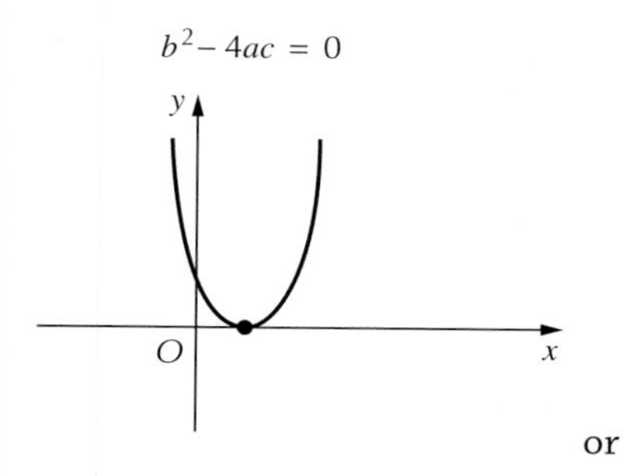

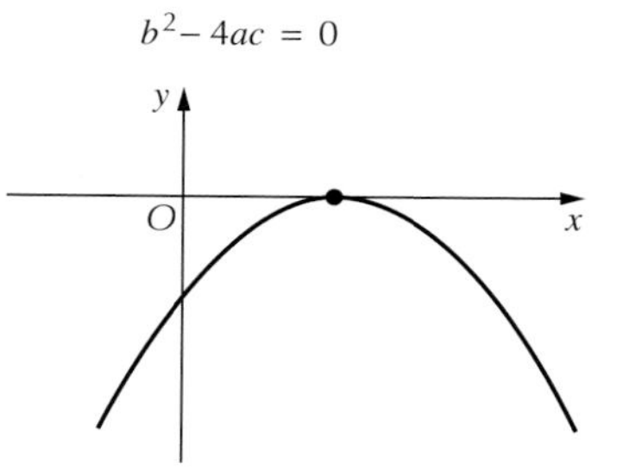

or

- No real roots, i.e. $b^2 - 4ac < 0$

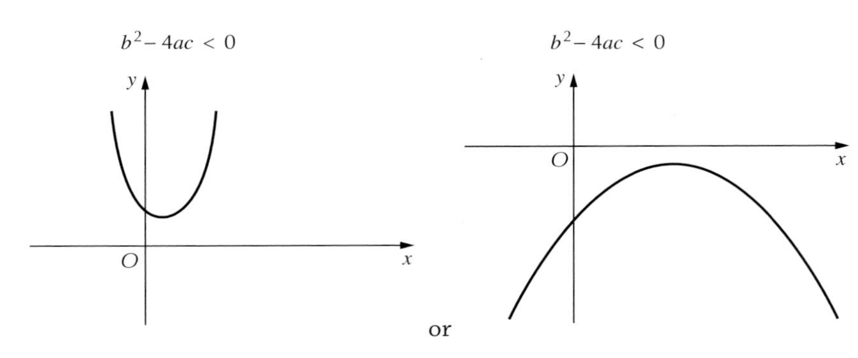

It is also worth remembering that for real roots (equal or distinct) the discriminant will be greater than or equal to zero, i.e. $b^2 - 4ac \geqslant 0$.

Also, if the value of $b^2 - 4ac$ turns out to be a square number, then the roots will be *rational*; otherwise the roots will be *irrational*. See Chapter 6 (Essential skills) for more on square numbers and the glossary for a definition of **rational** and **irrational**.

Example 12.5 

Q Functions f and g are defined on suitable **domains** by

$$f(x) = x(x-1) + q \text{ and } g(x) = x + 3.$$

(a) Find an expression for $f(g(x))$. **2**

(b) Hence, find the value of q such that the equation $f(g(x)) = 0$ has equal roots. **4**

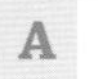

A (a)

$f(g(x))$

$= f(x + 3)$ (✓)

$= (x + 3)(x + 3 - 1) + q$ (✓)

Hint

See Chapter 10 (Functions and graphs) for more on **composite functions.**

Replace every x in the f function with the entire g function.

(b)

$f(g(x)) = 0$

$(x + 3)(x + 2) + q = 0$

$x^2 + 5x + 6 + q = 0$ (✓)

$a = 1$

$b = 5$

$c = (6 + q)$

For equal roots $b^2 - 4ac = 0$

$5^2 - 4(1)(6 + q) = 0$(✓)

$25 - 4(6 + q) = 0$

$25 - 24 - 4q = 0$ (✓)

$1 - 4q = 0$

$-4q = -1$

$q = \frac{1}{4}$ (✓)

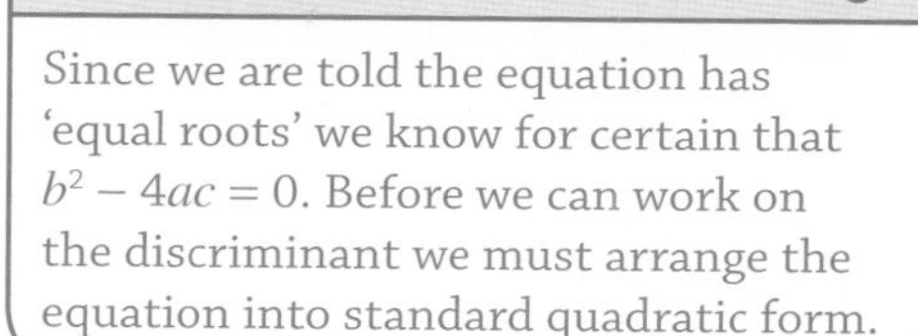

Since we are told the equation has 'equal roots' we know for certain that $b^2 - 4ac = 0$. Before we can work on the discriminant we must arrange the equation into standard quadratic form.

Start to simplify the expression for $f(g(x))$.

Remove the brackets and write in standard quadratic form.

Prepare to use the discriminant by writing down the coefficients of x^2 and x, and the number term. The number term here is $6 + q$ since q is just another number.

Substitute into the discriminant ($b^2 - 4ac$). The bracket around $6 + q$ is essential. Don't try to evaluate anything at this stage.

Simplify everything and take care with the signs when multiplying the -4 into the bracket.

Solve for q, taking care with the negatives.

Example 12.6 

Q

Hint

This question is very similar to the previous example except for the fact that we are working with no real roots and the discriminant $b^2 - 4ac < 0$ will create a quadratic inequality that can be tricky to solve. The word **'range'** tells you that there won't be a single numerical answer for p. Don't worry too much about '$p \in \mathbb{R}$', this is notation to tell you that p is a real number and doesn't affect how you answer the question. See the glossary for more on this notation.

Given that $2x^2 + px + p + 6 = 0$ has no real roots, find the range of values for p, where $p \in \mathbb{R}$. **4**

A

$2x^2 + px + p + 6 = 0$

$a = 2$

$b = p$

$c = p + 6$

> Prepare to use the discriminant by writing down the coefficients of x^2 and x, and the number term. The coefficient of x is p and the number term is $p + 6$ since p is just another number.

For no real roots, $b^2 - 4ac < 0$ (✓)

$p^2 - 4(2)(p + 6) < 0$

> Substitute into the discriminant ($b^2 - 4ac$). The bracket around $p + 6$ is essential. Don't attempt to evaluate anything at this stage.

$p^2 - 8(p + 6) < 0$

$p^2 - 8p - 48 < 0$ (✓)

> Simplify everything before multiplying into the bracket.

$(p - 12)(p + 4) < 0$ (✓)

> Factorise the quadratic but take care, this is a quadratic inequality and must not be made equal to 0.

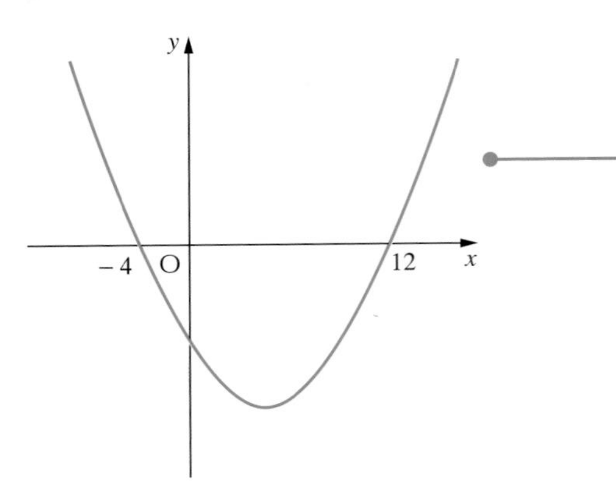

> To solve the quadratic inequality you can make a sketch of the quadratic and examine it to see where $p^2 - 8p - 48 < 0$, i.e. look for the parts of the graph that are below the x-axis. We know from the factors that when $p^2 - 8p - 48 = 0$, the roots are at -4 and 12. Since the coefficient of p^2 is positive, the graph will have a ∪ shape.

$-4 < p < 12$ (✓)

> Write down the range of values of p for which the graph is below the x-axis.

Trigonometric equations

Before you start this chapter you should be able to:

- Solve **linear** trigonometric equations using degrees,
  e.g. Solve $7 \sin x^\circ - 2 = 3,\ 0 \le x \le 360$. See Chapter 6 (Essential skills) for more on solving trigonometric equations in degrees.
- Use **exact value ratios** for sine, cosine and **tangent**, e.g. $\sin 30^\circ = \frac{1}{2}$, $\cos 45^\circ = \frac{1}{\sqrt{2}}$ etc. See Chapter 9 (Manipulating trigonometric **expressions**) for more on working with exact values.
- Work with **radian** equivalents of 30°, 45°, 60° and 90°, e.g. $30^\circ = \frac{\pi}{6}$ etc. See Chapter 9 (Manipulating trigonometric expressions) for more on working with radians.
- Use the identity $\cos^2 A + \sin^2 A = 1$.
- Use the addition and double angle formulae. See Chapter 9 (Manipulating trigonometric expressions) for more on the addition and double angle formulae.
- Convert $a \cos x + b \sin x$ to the forms $k \cos (x \pm \alpha)$ or $k \sin (x \pm \alpha)$. See Chapter 9 (Manipulating trigonometric expressions) for more on writing a sum or difference of trig terms as a single trig **function**.
- Sketch or recognise graphs of the form $y = a \cos bx + c$ and $y = a \cos (x + d) + c$ and the sine equivalents. See Chapter 6 (Essential skills) for more on working with the graphs of trig functions.

This chapter covers:

- Solving linear and quadratic trigonometric equations in degrees or radians, including those involving trigonometric formulae, over a given interval.

Solving a trigonometric equation involving a double angle in degrees

Look out for double angles, e.g. sin 2A or cos 2A and remember that there are substitutions available for them on your formula sheet. Choose the substitution for cos 2A carefully as it will affect how you answer the rest of the question. Some of these equations will require you to work in radians and many will be non-calculator, meaning you will have to use exact values.

- $\sin 2A = 2\sin A \cos A$
- $\cos 2A = \cos^2 A - \sin^2 A$
- $\cos 2A = 2\cos^2 A - 1$
- $\cos 2A = 1 - 2\sin^2 A$

Example 13.1 

Q Solve the equation $\sin 2x° = 6\cos x°$ for $0 \le x \le 360$. **4**

A

$\sin 2x° = 6\cos x°$

$2\sin x° \cos x° = 6\cos x°$ (✓)

Use the double angle substitution $\sin 2A = 2\sin A \cos A$, but remember to use x and not A.

$2\sin x° \cos x° - 6\cos x° = 0$

Rearrange to the form ... = 0.

$2\cos x°(\sin x° - 3) = 0$ (✓)

Factorise by taking out the highest common **factor**.

$2\cos x° = 0$ and $\sin x° - 3 = 0$

$\cos x° = 0$ and $\sin x° = 3$ (✓)

Set each factor equal to zero.

$x = 90°, 270°$ no solution since $\sin x° = 3 > 1$ (✓)

Solve each factor separately using your knowledge of the graphs of $y = \cos x°$ and $y = \sin x°$. You must explain why there is no solution to $\sin x° = 3$ before you can be awarded the last mark.

Solving a quadratic trigonometric equation in degrees involving a double angle over a given interval

In this type of question you will be required to write the double angle, e.g. $\cos 2x$, in terms of either $\sin^2 x$ or $\cos^2 x$. This will result in a quadratic trig equation, usually a trinomial (three terms), which you will need to factorise before it can be solved. The aim is to create a quadratic trig equation in either $\sin x$ or $\cos x$ so the substitution should be chosen carefully. If the given equation already contains $\cos x$ you should replace $\cos 2x$ with $2\cos^2 x - 1$ and if it already contains $\sin x$ you should replace $\cos 2x$ with $1 - 2\sin^2 x$.

Example 13.2 

Solve $\cos 2x° - 3\cos x° + 2 = 0$ for $0 \leq x < 360$. **5**

$\cos 2x° - 3\cos x° + 2 = 0$

$2\cos^2 x° - 1 - 3\cos x° + 2 = 0$ (✓)

$2\cos^2 x° - 3\cos x° + 1 = 0$ (✓)

$(2\cos^2 x° - 1)(\cos x° - 1) = 0$ (✓)

$2\cos x° - 1 = 0$ and $\cos x° - 1 = 0$

$2\cos x° = 1$

$\cos x° = \frac{1}{2}$ and $\cos x° = 1$ (✓)

$x = 60°, 300°$ and $x = 0°$ (✓)

Hint

Look closely at the given **domain** ($0 \leq x < 360$). In this question your answers could include 0, but not 360.

You want an equation with only $\cos x$ terms, this means you should use the double angle substitution $\cos 2A = 2\cos^2 A - 1$.

Simplify and rearrange into the form ... = 0. You must include the '= 0' to get all the marks.

Factorise this quadratic trig equation in the same way as a normal quadratic. It can be helpful when factorising to ignore the trig terms and write it out as $2x^2 - 3x + 1 = (2x - 1)(x - 1)$. Just remember to put the trig terms back in when you have found the factors.

Set each factor equal to zero.

Reduce the equations to $\cos x° = ...$

Solve each factor separately using your knowledge of exact values and of the graph of $y = \cos x°$. If you include 360° in your answers you won't be awarded the last mark as 360° is outside the given domain. See Chapter 9 (Manipulating trigonometric expressions) for more on exact values.

Solving a quadratic trigonometric equation involving a double angle over a given interval using radians

Trigonometric equations expressed in radians are solved using the same methods as those expressed in degrees. However, it is important to remember that if a question using radians asks for exact solutions, then you should give the solutions as fractions or multiples of π. See Chapter 9 (Manipulating trigonometric expressions) for more on working with radians.

Example 13.3 

Q Solve the equation $3\cos 2x + 10\sin x + 1 = 0$ for $0 \le x \le 2\pi$, correct to 2 decimal places. **5**

Hint

The $0 \le x \le 2\pi$ tells you that you should be working in radians and that any solutions you give should be between 0 and 2π. See Chapter 9 (Manipulating trigonometric expressions) for more on working with radians.

A $3\cos 2x + 10\sin x + 1 = 0$

$3(1 - 2\sin^2 x) + 10\sin x + 1 = 0$ (✓)

Since the given equation is a trinomial (three terms) and because you want to create an equation with only $\sin x$ terms, you should use the double angle substitution $\cos 2x = 1 - 2\sin^2 x$. It will have to be multiplied by 3 so make sure you put it in brackets.

$3 - 6\sin^2 x + 10\sin x + 1 = 0$

Take care multiplying out the brackets and don't forget to multiply both terms in $(1 - 2\sin^2 x)$ by 3.

$-6\sin^2 x + 10\sin x + 4 = 0$ (✓)

$3\sin^2 x - 5\sin x - 2 = 0$

Simplify by collecting like terms. You can divide throughout by -2 here if you want. This won't affect your solutions but will make the numbers smaller and the next factorising step a little easier.

$(3\sin x + 1)(\sin x - 2) = 0$ (✓)

Factorise this quadratic trig equation in the same way as a normal quadratic. It can be helpful when factorising to ignore the trig terms and write it out as $3x^2 - 5x - 2 = (3x + 1)(x - 2)$. Just remember to put the trig terms back in when you have found the factors.

$3\sin x + 1 = 0$ and $\sin x - 2 = 0$

Set each factor equal to zero.

$3\sin x = -1$

$\sin x = -\frac{1}{3}$ and $\sin x = 2$ (✓)

Reduce the equations to $\sin x = ...$

1st quadrant angle $= \sin^{-1}\left(\frac{1}{3}\right) = 0.339...$

Ignore the (–) and find the solution to $\sin^{-1}\left(\frac{1}{3}\right)$. Make sure you change your calculator to radians mode.

no solution since $\sin x = 2 > 1$

You must explain why there is no solution to $\sin x = 2$ before you can be awarded the last mark.

S	A
✓ T	C ✓
$\pi + ...$	$2\pi - ...$

$x = \pi + 0.339...$ and $x = 2\pi - 0.339...$

Use the 1st quadrant angle and the four-quadrant diagram to find the solutions to $x = \sin^{-1}\left(-\frac{1}{3}\right)$. The solutions will be in the 3rd and 4th quadrants, i.e. the quadrants where sine is negative.

$x = 3.48, 5.94$ (to 2 decimal places) (✓)

Make sure you leave your answer rounded as specified in the question.

Calculus: Differentiation

Before you start this chapter you should be able to:

- Calculate the **gradient** of a straight line given suitable information (Chapter 16)
- Apply the laws of indices to multiplying out brackets and simplifying (Chapter 6)
- Solve **linear** and quadratic equations (Chapter 6)
- Simplify numerical and algebraic fractions (Chapter 9)
- Convert between **radians** and degrees (Chapter 9)
- Find the **exact value** of a simple trigonometric **expression** (Chapter 9)
- Determine a **composite function** $f(g(x))$ when given $f(x)$ and $g(x)$ (Chapter 10).

This chapter covers:

- Differentiating an algebraic **function** which can be simplified to an expression in powers of x
- Differentiating $k \sin x$, $k \cos x$
- Differentiating a composite function using the **chain rule**
- Using **differentiation** to find the equation of a **tangent** to a curve at a given point
- Determining when a function is **strictly increasing/decreasing**
- Sketching the graph of a function by determining the nature of **stationary points** and **intersections** with the axes.

Differentiating an algebraic function that can be simplified to an expression in powers of x

Questions which ask you to: 'differentiate'; 'find the **derivative**'; find '$f'(x)$'; 'find $\frac{dy}{dx}$'; or find a '**rate of change**', all require the same process. Before differentiating you must prepare the function by replacing square **roots** (and other roots) with fractional indices and by writing terms in the form ax^n, i.e. no x terms on the denominator, even if it means negative indices. Most marks are lost in these questions due to errors with indices, negatives and fractions. See Chapter 6 (Essential skills) for more on working with negatives, fractions and indices.

Hint

The process for differentiating is to '*multiply* by the **index** *then subtract* one from the index'. In general this means that differentiating $y = ax^n$ gives $\frac{dy}{dx} = anx^{n-1}$.

Example 14.1 

Given that $f(x) = \sqrt{x} + \frac{2}{x^2}$, find $f'(4)$. **5**

$f(x) = \sqrt{x} + \frac{2}{x^2}$

$f(x) = x^{\frac{1}{2}} + 2x^{-2}$ (✓)(✓)

Remove root signs and write terms in the form ax^n. One term correct gets the first mark, completing the preparation gets the second mark.

$f'(x) = \frac{1}{2}x^{-\frac{1}{2}} - 4x^{-3}$ (✓)(✓)

Differentiate one term at a time by multiplying by the index then subtracting one from the index. Again, one term correct gets a mark and completing the differentiation gets another mark.

$f'(x) = \frac{1}{2x^{\frac{1}{2}}} - \frac{4}{x^3}$

Write with positive indices. Remember, only the term with the index (usually x) will move from the numerator to the denominator of the fraction. Numbers in the numerator or denominator almost always stay where they are.

$f'(x) = \frac{1}{2\sqrt{x}} - \frac{4}{x^3}$

Replacing fractional indices with roots will make the next step much easier.

$f'(4) = \frac{1}{2\sqrt{4}} - \frac{4}{4^3}$

Substitute the value of 4 into the derivative in place of x.

$f'(4) = \frac{1}{4} - \frac{1}{16} = \frac{3}{16}$ (✓)

Evaluate $f'(4)$. Think of $\frac{1}{4} - \frac{1}{16}$ as $\frac{4}{16} - \frac{1}{16}$.

Hint

The strategy for questions like this is: prepare for differentiation → differentiate → write with positive indices (and root signs) → substitute → evaluate. Positive indices and root signs make the evaluation step much easier.

Use differentiation to find the equation of a tangent to a curve at a given point

To answer this type of question you need to know that the derivative of a function represents the gradient of the curve at a point. You will have to use your straight line knowledge to complete this type of question. See Chapter 16 (The straight line) for more on working with gradients and equations of straight lines.

Example 14.2 

Q The point P(x, y) lies on the curve with equation $y = 6x^2 - x^3$.

(a) Find the value of x for which the gradient of the tangent at P is 12. **5**

(b) Hence find the equation of the tangent at P. **2**

Keyword

The word 'gradient' in this question is your link to differentiation.

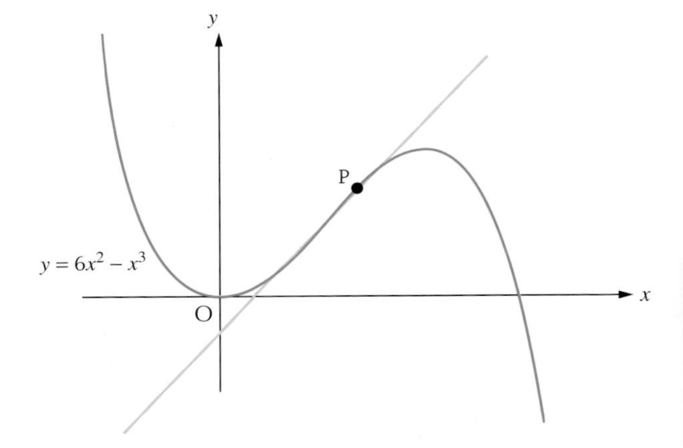

Hint

This diagram shows the curve $y = 6x^2 - x^3$ and the tangent whose equation you have to find. The diagram lets you see that the curve and the tangent have the same gradient at the point P. The first step in this question will be to find the x-coordinate of P by differentiating the function and making the result equal to 12.

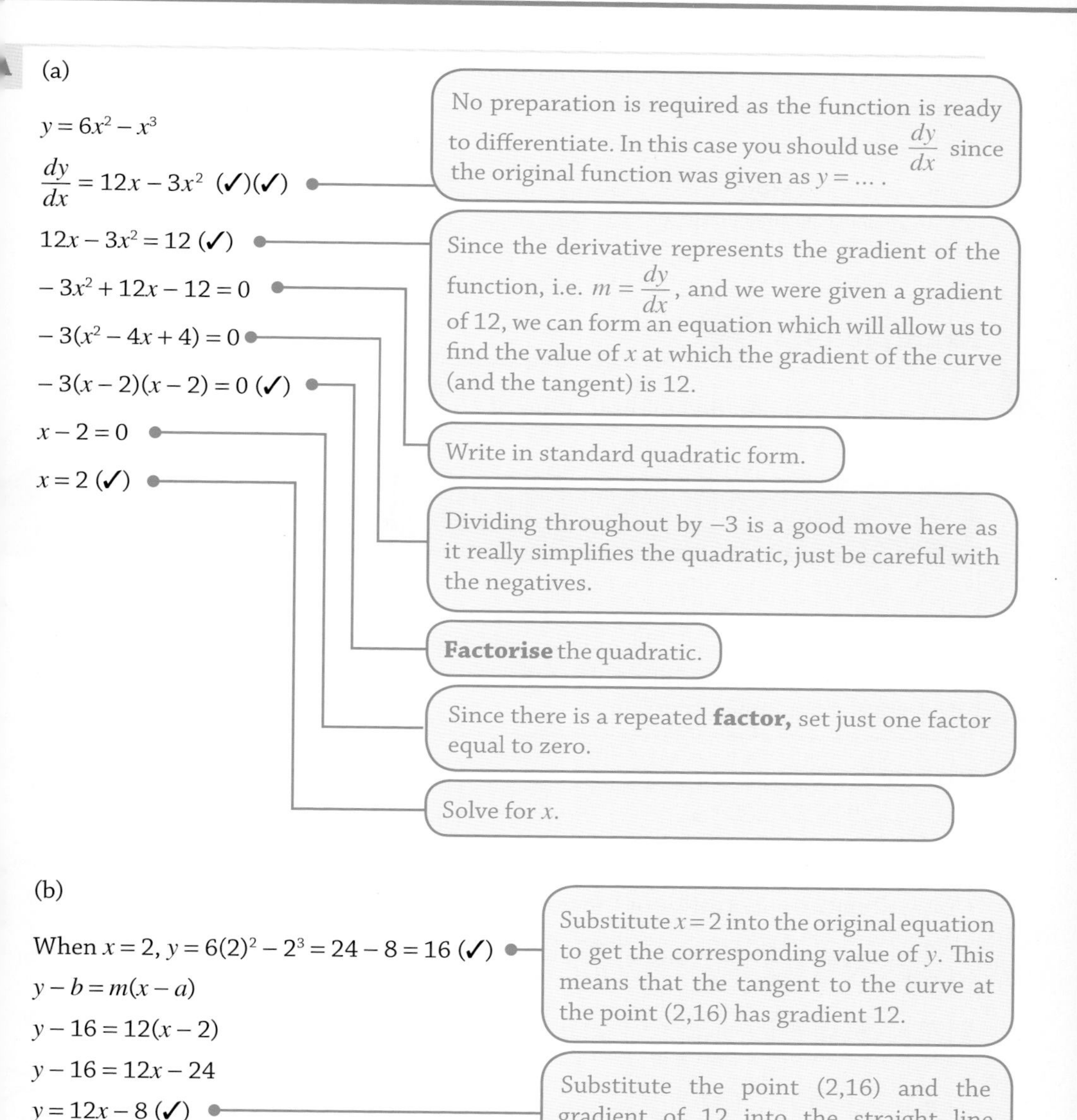

Differentiating a composite function using the chain rule

You are given some standard results for differentiating $\sin x$ and $\cos x$ in your formula sheet, although we suggest you remember this diagram as it makes differentiating − sin and − cos easier. See Chapter 7 (Formulae) for the formulae given to you in your exam, plus some other useful formulae you should be familiar with. This diagram will also help you to **integrate** trigonometric functions. See Chapter 15 (Integration) for more on integrating trigonometric functions.

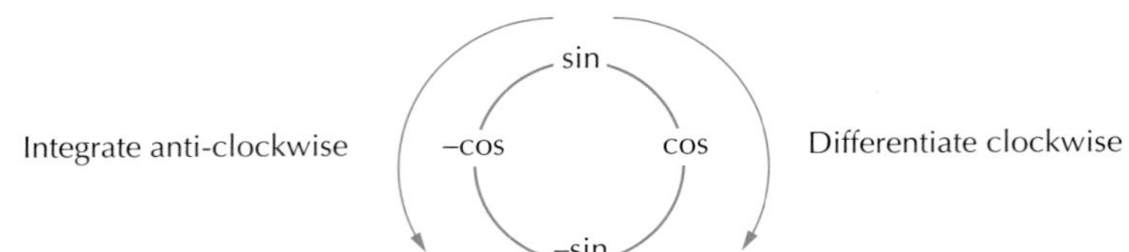

Example 14.3 

Q

If $f(x) = \cos 2x - 3\sin 4x$, find the exact value of $f'\left(\frac{\pi}{6}\right)$. **4**

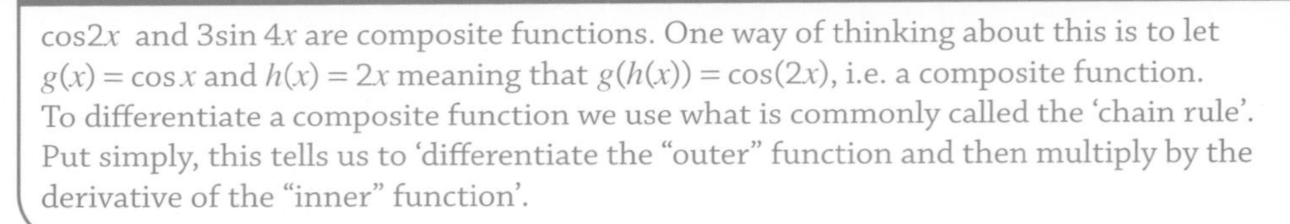

Hint

$\cos 2x$ and $3\sin 4x$ are composite functions. One way of thinking about this is to let $g(x) = \cos x$ and $h(x) = 2x$ meaning that $g(h(x)) = \cos(2x)$, i.e. a composite function. To differentiate a composite function we use what is commonly called the 'chain rule'. Put simply, this tells us to 'differentiate the "outer" function and then multiply by the derivative of the "inner" function'.

A

$f(x) = \cos 2x - 3\sin 4x$

$f(x) = \cos(2x) - 3\sin(4x)$ — Putting brackets around the 'inner' functions helps when using the chain rule.

$f'(x) = -2\sin(2x)\ldots$ (✓) — In the term $\cos(2x)$, the 'outer function' is cos(...) and the 'inner' function is $2x$. The cos(...) differentiates to –sin(...) then we multiply this result by the derivative of $2x$(which is 2), all the time keeping the function inside the brackets the same.

$f'(x) = -2\sin(2x) - 12\cos(4x)$ (✓) — In the term $-3\sin(4x)$, the 'outer function' is –3sin(...) and the inner function is $4x$. The –3sin(...) differentiates to –3cos(...) then we multiply this result by the derivative of $4x$ (which is 4), again keeping the function inside the brackets the same.

$f'\left(\frac{\pi}{6}\right) = -2\sin\left(2\left(\frac{\pi}{6}\right)\right) - 12\cos\left(4\left(\frac{\pi}{6}\right)\right)$ (✓) — Substitute $x = \frac{\pi}{6}$ into the derivative.

$f'\left(\frac{\pi}{6}\right) = -2\sin\left(\frac{\pi}{3}\right) - 12\cos\left(\frac{2\pi}{3}\right)$ — Simplify.

$f'\left(\frac{\pi}{6}\right) = -2\left(\frac{\sqrt{3}}{2}\right) - 12\left(-\frac{1}{2}\right)$ — Use your knowledge of exact values and the four-quadrant diagram to obtain the exact values of $\sin\left(\frac{\pi}{3}\right)$ and $\cos\left(\frac{2\pi}{3}\right)$, then simplify your answer.

$f'\left(\frac{\pi}{6}\right) = 6 - \sqrt{3}$ (✓)

Sketching the graph of a function by determining stationary points and intercepts with the axes

Most of these questions are very similar and you should be able to pick up lots of marks if you practise and remember the steps.

Hint

The general process is: differentiate → set the derivative equal to zero → solve and find the stationary points (SPs) → create a **nature table** → determine the nature and coordinates of the SPs → find where the graph crosses the x and y-axes → sketch the graph.

Example 14.4

A curve has equation $y = 3x^2 - x^3$.

(a) Find the coordinates of the stationary points on this curve and determine their nature. **6**

(b) State the coordinates of the points where the curve meets the coordinate axes and sketch the curve. **2**

Keyword

The key words here are 'stationary points' and 'nature'. When you see them you should think differentiation and nature table.

(a)

$y = 3x^2 - x^3$

$\frac{dy}{dx} = 6x - 3x^2$ (✓)

Differentiate each term by 'multiplying by the power then subtracting one from the power'.

S.P.s occur when $\frac{dy}{dx} = 0$

$6x - 3x^2 = 0$ (✓)

Set the derivative to zero and solving will allow you to find the x-coordinates of the points where the gradient of the graph is zero, i.e. the stationary points.

$3x(2 - x) = 0$

Always check for a common factor before doing any other type of factorising.

$3x = 0$ and $2 - x = 0$

Set every factor equal to zero.

$x = 0$ and $x = 2$ (✓)

Solve to find the x-coordinates of the stationary points.

When $x = 0$,

$y = 3(0)^2 - (0)^3 = 0 \quad (0, 0)$

When $x = 2$,

$y = 3(2)^2 - (2)^3 = 12 - 8 = 4$

$(2, 4)$ (✓)

Substitute your x values into the original function (not the derivative) to find the corresponding y-values. These are the coordinates of the stationary points.

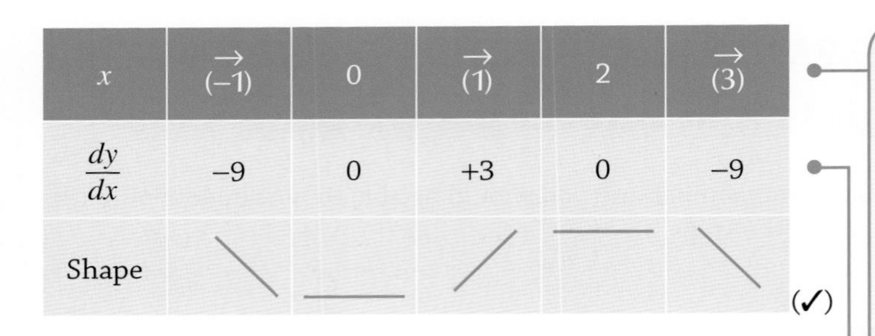

x	$\overrightarrow{(-1)}$	0	$\overrightarrow{(1)}$	2	$\overrightarrow{(3)}$
$\frac{dy}{dx}$	-9	0	$+3$	0	-9
Shape	\	_	/	‾	\

(✓)

Use a nature table to investigate the gradient of the curve on either side of the stationary points. Knowing whether the gradient is positive or negative will tell you the shape of the curve.

Substitute sensible values into your derivative and indicate whether the result is positive or negative. You already know the gradient will be zero at the stationary points.

Hint

Make sure you include all the information shown above in your nature table, otherwise you risk losing marks. Some functions involve a variable (letter) other than x, in these cases you must make sure you use the variable you are given in the question.

Minimum turning point at (0, 0)

Maximum turning point at (2, 4) (✓)

Use the table to help you determine the nature of the stationary points and clearly communicate your results, remembering to give coordinates.

(b)

$y = 3x^2 - x^3$

$3x^2 - x^3 = 0$

$x^2(3 - x) = 0$

$x^2 = 0$ and $3 - x = 0$

$x = 0$ and $x = 3$

Crosses x-axis at (0, 0) and (3, 0)

$y = 3x^2 - x^3$

$y = 3(0)^2 - (0)^3 = 0$

Crosses y-axis at (0, 0)

Set the original function equal to zero and solve to find where it crosses the x-axis. Sometimes this step might need synthetic division, but not always.

This function can be fully factorised by taking out a common factor.

Set both factors equal to zero, then solve to find where the function crosses the x-axis.

Substitute $x = 0$ into the original function to find where the graph crosses the y-axis.

y

(2, 4)

O

3

x

Make a small and neat sketch of the curve using the shape from your nature table. You must clearly indicate the position of the **turning points** and the points where the graph crosses **both** axes.

Marks for this question are awarded for:

- Finding where the curve crosses the axes (✓)
- Making an accurate and labelled sketch (✓)

Determining when a function is strictly increasing/decreasing

This is a differentiation question. When talking about functions increasing or decreasing, we are referring to the gradient of the function being positive (increasing) or negative (decreasing). Remember, if you want to work with *gradient*, find the *derivative*!

Example 14.5 

A function f is defined by $f(x) = \frac{1}{3}x^3 + 2x^2 - 5x + 1$.

(a) Determine whether the function is increasing, decreasing or stationary when $x = 3$. **3**

(b) Find the **range** of values of x for which $f(x)$ is decreasing. **3**

Keyword

Words like 'increasing', 'decreasing' and 'stationary' are referring to gradient. Differentiation will be required here.

Hint

A common mistake here is to substitute '3' into the original function. This will give the value of the function at the point when $x = 3$ but tells us nothing about the gradient of the function.

(a)

$f(x) = \frac{1}{3}x^3 + 2x^2 - 5x + 1$

$f'(x) = x^2 + 4x - 5$ (✓)

$f'(3) = 3^2 + 4(3) - 5$

$f'(3) = 9 + 12 - 5$

$f'(3) = 16$ (✓)

Since $f'(3) > 0$ the function is increasing when $x = 3$ (✓)

To decide whether the function is increasing, decreasing or stationary you need to examine whether the gradient is positive, negative or zero. To obtain the gradient you will first have to differentiate the function.

Substitute '3' into the derivative to find the gradient of the function at the point where $x = 3$.

Evaluate the gradient; it will be positive, negative or zero.

Communicate your final answer clearly. Use the greater than sign (>) and not the greater than or equal to sign (≥) so you include only positive gradient.

Hint

There are different ways to tackle this question but a nature table is a good choice as you should be familiar with constructing one and it shows clearly the range of values for which the function is increasing and/or decreasing.

(b)

$f'(x) = x^2 + 4x - 5$

S.P.s occur when $f'(x) = 0$

Set $f'(x) = 0$ and solve to find the roots.

$x^2 + 4x - 5 = 0$

$(x - 1)(x + 5) = 0$

$x = 1$ and $x = -5$ (✓)

x	$\overrightarrow{(-6)}$	-5	$\overrightarrow{(0)}$	1	$\overrightarrow{(2)}$
$f'(x)$	7	0	−5	0	7
Shape	/	—	\	—	/

(✓)

The function is decreasing when $f'(x) < 0$ (✓).

$f(x)$ is decreasing for $-5 < x < 1$ (✓)

You must communicate that you know that the derivative is negative when the function is decreasing.

From the nature table you can see that $f'(x)$ is less than zero, i.e. below the x-axis, between −5 and 1.

Integration

Before you start this chapter you should be able to:

- Differentiate algebraic, trigonometric and **composite functions** (Chapter 14).

This chapter covers:

- **Integrating** (and evaluating) an algebraic **function** which can be simplified to an **expression** in powers of x
- Integrating (and evaluating) functions of the form $f(x) = (px + q)^n$
- Integrating (and evaluating) functions of the form $f(x) = p\sin x$ and $f(x) = p\cos x$, $f(x) = p\sin(qx + r)$ and $f(x) = p\cos(qx + r)$
- Solving differential equations of the form $\dfrac{dy}{dx} = f(x)$.

Integrating an algebraic function which can be simplified to an expression in powers of x

Before integrating you will often have to prepare the function by replacing square **roots** (and other roots) with fractional indices and by writing terms in the form ax^n, i.e. no x terms in the denominator, even if this means using negative indices. Most marks are lost in these questions due to errors with indices, negatives and fractions. See Chapter 6 (Essential skills) for more on working with negatives, fractions and indices.

All **integrals** will have dx (or similar) written after them. Make sure you include this in any integrals you write down. This is not part of the function and won't be involved in your working but it must be there because it tells us that we are integrating *with respect to* x. When you integrate the dx will disappear and the **constant of integration** ($+C$) appears.

Hint

The process for integrating is to '*add* one to the **index** *then divide* by the new index'. In general if $\dfrac{dy}{dx} = kx^n$ then $y = \dfrac{kx^{n+1}}{n+1} + C$.
Don't confuse this with **differentiation**, which is the opposite process.

Example 15.1 

Q Find $\int \frac{3x^3+1}{2x^2}dx,\ x \neq 0.$ **4**

A

$$\int \frac{3x^3+1}{2x^2}\,dx$$

$$= \int\left(\frac{3x^3}{2x^2}+\frac{1}{2x^2}\right)dx$$

The function needs to be prepared for integration. The first step here is to write as two separate fractions.

$$= \int\left(\frac{3x}{2}+\frac{x^{-2}}{2}\right)dx \ (\checkmark)$$

Use your knowledge of indices to simplify the first fraction and move the x^2 in the second fraction to the numerator. The function is now ready to be integrated.

$$= \frac{3x^2}{2(2)}+\frac{x^{-1}}{2(-1)}+C \ (\checkmark)(\checkmark)$$

The procedure for integration is 'add one to the index then divide by the new index'. Don't forget the $+C$, if it is missing you will lose a mark.

$$= \frac{3x^2}{4}-\frac{x^{-1}}{2}+C \ (\checkmark)$$

Simplify. In this example the -1 which was in the denominator of the second fraction makes the whole fraction negative.

$$= \frac{3x^2}{4}-\frac{1}{2x}+C$$

Get into the habit of leaving your answers with positive indices. This will make longer questions where you are required to substitute values for x more straightforward. See Chapter 6 (Essential skills) for more on working with indices.

Hint

This is an ***indefinite* integral**, i.e. there is not a numerical answer and your result will be another function. Always remember the '$+C$' when integrating indefinite integrals. In this question the $x \neq 0$ has to be there because it prevents a division by zero. However, it has no effect on the way you should answer the question.

Integrate and evaluate functions of the form $f(x) = (px + q)^n$

Sometimes called the '**chain rule** for integration' composite functions of this type need to be integrated using the following rule:

$$\int (px+q)^n \, dx = \frac{(px+q)^{n+1}}{p(n+1)} + C$$

Here the 'outer' function is $(\ldots)^n$ and the inner function is $(px + q)$. You should integrate the 'outer' function as usual, i.e. add one to the index then divide by the new index, then also divide by the ***derivative*** of the inner function.

Example 15.2 

Find $\int_0^2 \sqrt{4x+1} \, dx$. **5**

> **Hint**
>
> This type of integral is called a ***definite* integral** and will eventually work out to a numerical value. There is no need to include a '$+C$' in your answer.

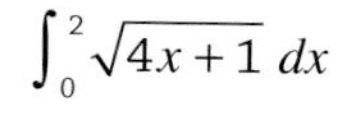

$$\int_0^2 \sqrt{4x+1} \, dx$$

$$= \int_0^2 (4x+1)^{\frac{1}{2}} \, dx \quad (\checkmark)$$

Prepare the function for integrating by rewriting root signs as fractional indices. This is a composite function and we will need to use the chain rule for integration. Here the *inner* function is $4x + 1$ and the *outer* function is $(\ldots)^{\frac{1}{2}}$.

$$= \left[\frac{(4x+1)^{\frac{3}{2}}}{\frac{3}{2}(4)} \right]_0^2 \quad (\checkmark)(\checkmark)$$

Add one to the power, i.e. $\frac{1}{2}$ becomes $\frac{3}{2}$, divide by this new power, then divide by the *derivative* of the *inner* function, i.e. the derivative of $4x + 1$ is 4.

$$= \left[\frac{\sqrt{(4x+1)^3}}{6} \right]_0^2$$

Putting the roots signs back in and simplifying the denominator makes it much easier to evaluate the integral.

$$= \left(\frac{\sqrt{(4(2)+1)^3}}{6} \right) - \left(\frac{\sqrt{(4(0)+1)^3}}{6} \right) \quad (\checkmark)$$

Substitute the limits into the integral in the order 'upper' – 'lower'. A lower limit of zero will often evaluate to zero, although this is not always the case, especially with trigonometric functions, so be careful.

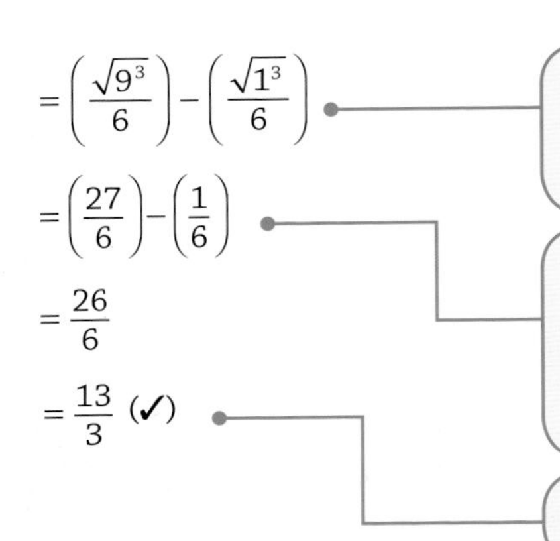

$$= \left(\frac{\sqrt{9^3}}{6}\right) - \left(\frac{\sqrt{1^3}}{6}\right)$$

$$= \left(\frac{27}{6}\right) - \left(\frac{1}{6}\right)$$

$$= \frac{26}{6}$$

$$= \frac{13}{3} \ (\checkmark)$$

Take it one step at a time. Carefully work out what is under the root signs. Thinking of $\sqrt{9^3}$ as $\left(\sqrt{9}\right)^3$ will make the next step easier.

To evaluate $\sqrt{9^3}$ first find the square root of 9, i.e. 3, then cube the result, i.e. $3^3 = 27$ See Chapter 6 for more on working with commonly occurring powers and roots.

Remember to fully simplify your answer.

Integrating (and evaluating) functions of the form $f(x) = p\sin x$ and $f(x) = p\cos x$, $f(x) = p\sin(qx + r)$ and $f(x) = p\cos(qx + r)$

There are four rules you should be familiar with when integrating trigonometric functions. They can be written in general as follows:

Rule	Example
$\int p\cos x\,dx = p\sin x + C$	$\int 6\cos x\,dx = 6\sin x + C$
$\int p\sin x\,dx = -p\cos x + C$	$\int 3\sin x\,dx = -3\cos x + C$
$\int p\cos(qx + r)\,dx = \frac{p\sin(qx + r)}{q}$	$\int 5\cos(2x - 3)\,dx = \frac{5\sin(2x - 3)}{2} + C$
$\int p\sin(qx + r)\,dx = -\frac{p\cos(qx + r)}{q}$	$\int 4\sin(7x + 3)\,dx = -\frac{4\cos(7x + 3)}{7} + C$

Example 15.3

Evaluate $\int_0^{\frac{\pi}{4}} 6\cos 2x \, dx$ **4**

Keyword

The word 'evaluate' tells you that your final answer should be a number.

Hint

Use the chain rule for integration to integrate this composite function. Here the inner function is $2x$ and the outer function is 6 cos (...). Use the table of integrals on your formula sheet, or the circle diagram in Chapter 7 (Formulae) to help you integrate any trigonometric terms.

$$\int_0^{\frac{\pi}{4}} 6\cos 2x \, dx$$

$$= \left[\frac{6\sin 2x}{2}\right]_0^{\frac{\pi}{4}} \quad (\checkmark)(\checkmark)$$

Integrating the outer function, 6 cos (...) gives 6 sin (...), but don't forget to divide by the derivative of the inner function, i.e. the derivative of $2x$ is 2.

$$= [3\sin 2x]_0^{\frac{\pi}{4}}$$

Always fully simplify your integral before substituting in the limits.

$$= \left(3\sin 2\left(\frac{\pi}{4}\right)\right) - (3\sin 2(0)) \quad (\checkmark)$$

Take a line of working just to substitute in the limits. Trying to evaluate at the same time can lead to mistakes.

$$= \left(3\sin\frac{\pi}{2}\right) - (3\sin 0)$$

Simplify before evaluating.

$$= (3(1)) - (3(0))$$

$$= 3 \quad (\checkmark)$$

Use your knowledge of the graph of $y = \sin x$ to evaluate $\sin\frac{\pi}{2}$ and sin 0, i.e.

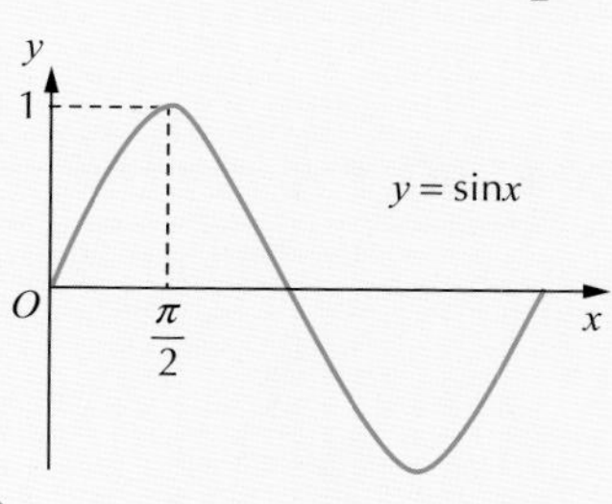

Solving differential equations of the form $\frac{dy}{dx} = f(x)$

It is important that you realise this question requires integration. Integration is sometimes called *anti-differentiation* because it is the 'reverse' of differentiating. See Chapter 14 (Differentiation) for more on differentiating functions. Since differentiating y gives $\frac{dy}{dx}$, integrating $\frac{dy}{dx}$ will get back to y. Don't forget to add the constant of integration ($+ C$) then use the given point to find the constant (C) before finally writing y in terms of x. In some questions you might be given a graph and might be required to determine a point on the graph before you can work out the constant. You would follow the same process if you were given $f'(x)$ and asked to find $f(x)$.

Example 15.4 

Q The curve $y = f(x)$ is such that $\frac{dy}{dx} = 4x - 6x^2$. The curve passes through the point $(-1, 9)$. Express y in terms of x. **4**

A

$\frac{dy}{dx} = 4x - 6x^2$

$y = \int 4x - 6x^2\, dx$ (✓)

Turn $\frac{dy}{dx}$ into y by integrating. Don't forget to include the dx.

$y = \frac{4x^2}{2} - \frac{6x^3}{3} + C$ (✓)

Integrate each term by adding one to the power then dividing by the new power. Remember to $+C$.

$y = 2x^2 - 2x^3 + C$

Simplify.

When $x = -1$, $y = 9$ so,

$9 = 2(-1)^2 - 2(-1)^3 + C$ (✓)

Substitute the given point into your result to find C.

$9 = 2 + 2 + C$

$C = 5$

$y = 2x^2 - 2x^3 + 5$ (✓)

To get the final mark you must write out the equation for y in full, including your value for C.

Unit 3: Applications

The straight line

Before you start this chapter you should be able to:

- Use the **gradient** formula $m = \dfrac{y_2 - y_1}{x_2 - x_1}$
- Use the formula $y - b = m(x - a)$ to find the equation of a straight line
- Identify the gradient and y – intercept from various forms of the equation of a straight line
- Find the midpoint of a line segment
- Calculate the distance between two points
- Find the point of **intersection** of two lines by solving two **linear** equations simultaneously (Chapter 6).

This chapter covers:

- Finding the equation of a line **parallel** to and a line perpendicular to a given line
- Using properties of **medians**, **altitudes** and perpendicular **bisectors**
- Using $m = \tan\theta$ to calculate a gradient or an angle
- Solving problems involving straight lines.

Find the equation of a line parallel to and a line perpendicular to a given line

In this type of question you will usually need to find the gradient of a line from its equation. This will most likely require you to change the subject of the given equation to y. Doing this will leave an equation in the form $y = mx + c$ and will allow you to 'read off' m, the gradient.

Example 16.1 

Find the equation of the line which passes through the point (–1, 3) and is perpendicular to the line with equation $4x + y - 1 = 0$. **3**

$4x + y - 1 = 0$

$y = -4x + 1$

Change the subject of the equation to y so you can read off the gradient.

$m = -4$ (✓)

When written in the form $y = mx + c$, the gradient of a straight line is the **coefficient** of x.

$m_{\perp} = \frac{1}{4}$ since $-4 \times \frac{1}{4} = -1$ (✓)

Find the perpendicular gradient using $m_{\perp} = -\frac{1}{m}$ and remember to justify your answer by writing that $m \times m_{\perp} = -1$.

$y - b = m(x - a)$

Use this formula to find the equation of a straight line when you know the gradient and a point on the line. You will have to remember this formula as it is not given to you in your examination.

$y - 3 = \frac{1}{4}(x - (-1))$

Substitute the perpendicular gradient and the given point, i.e. (–1, 3) in for a and b.

$4y - 12 = x + 1$

Remove any fractions by multiplying both sides of the equation by the denominator (4 in this case).

$4y = x + 13$ (✓)

Other (correct) versions of this equation are also acceptable, e.g. $4y - x = 13$ or $4y - x - 13 = 0$.

Hint

Parallel lines have equal gradients. **Perpendicular lines** have gradients that multiply to make –1. Often $m_{\perp}$ is used to indicate a perpendicular gradient.

Example 16.2

Q

(a) Find the equation of l_1, the perpendicular bisector of the line joining P(3, –3) to Q(–1, 9). **4**

(b) Find the equation of l_2 which is parallel to PQ and passes through R(1, –2). **2**

(c) Find the point of intersection of l_1 and l_2. **3**

(d) Hence find the shortest distance between PQ and l_2. **2**

Keyword

A 'perpendicular bisector' cuts another line in half and meets it at 90°. In this question you will usually have to find the gradient of the original line, a perpendicular gradient and also the midpoint of the original line. All this needs to be done before using $y - b = m(x - a)$ to find the equation of the perpendicular bisector.

A (a)

$$m_{PQ} = \frac{y_2 - y_1}{x_2 - x_1} = \frac{9-(-3)}{-1-3} = \frac{12}{-4} = -3 \ (\checkmark)$$

Use the gradient formula to calculate the gradient of the original line, taking care with negatives. You will have to memorise the gradient formula as it is **not** given to you in your examination.

$$m_{l_1} = \frac{1}{3} \quad \text{since} \quad -3 \times \frac{1}{3} = -1 \ (\checkmark)$$

Find the perpendicular gradient using $m_\perp = -\frac{1}{m}$ and remember to justify your answer by writing that $m \times m_\perp = -1$.

$$M\left(\frac{3+(-1)}{2}, \frac{-3+9}{2}\right)$$

M (1, 3) (✓)

Give the midpoint of PQ a letter and calculate it by finding the mean of the x and the mean of the y-coordinates, i.e. 'add them then half them'.

$$y - b = m(x - a)$$

Find the equation of line l_1.

$$y - 3 = \frac{1}{3}(x - 1)$$

Make sure you substitute in the *perpendicular* gradient and the mid-point, i.e. (1, 3) for a and b.

$$3y - 9 = x - 1$$

Remove fractions by multiplying both sides by any denominator(s).

$$3y = x + 8 \ (\checkmark)$$

It makes sense to fully simplify the equation now as it is often required later in the question. Other (correct) versions of the equation are also acceptable.

(b)

$m_{l_2} = m_{PQ} = -3$ (✓) — Use the fact that parallel lines have equal gradients to find the gradient of l_2.

$y - b = m(x - a)$ — Find the equation of line l_2.

$y - (-2) = -3(x - 1)$ — Substitute in the gradient and the given point, i.e. $(1, -2)$ for a and b.

$y + 2 = -3x + 3$

$y = -3x + 1$ (✓) — Simplify the equation. Other (correct) versions of the equation are also acceptable.

(c)

$3y = x + 8$

$3(-3x + 1) = x + 8$ (✓) — Use simultaneous equations to find where two lines intersect (meet). Since we already simplified the equations in parts (a) and (b), we can substitute to create an equation in x. The elimination method for simultaneous equations will also work here.

$-9x + 3 = x + 8$ — Remove the brackets.

$-10x = 5$ — Simplify.

$x = -\frac{1}{2}$ (✓) — Divide both sides of the equation by -10, remembering that $5 \div -10 = -\frac{1}{2}$.

When $x = -\frac{1}{2}$,

$y = -3\left(-\frac{1}{2}\right) + 1 = \frac{5}{2}$ — Substitute your answer for x into either of the original straight line equations to find the corresponding y value. Take care with the negatives and the fractions here. Do some extra working at the side if you want to. Thinking of 1 as $\frac{2}{2}$ will help.

The point of intersection of lines l_1 and l_2 is $\left(-\frac{1}{2}, \frac{5}{2}\right)$ (✓) — Make sure you communicate clearly what you have found and give your answer as coordinates.

(d)

Hint

The shortest distance between parallel lines is the *perpendicular* distance.

Keyword

The word 'hence' tells you that you should use your earlier results to help answer this part of the question.

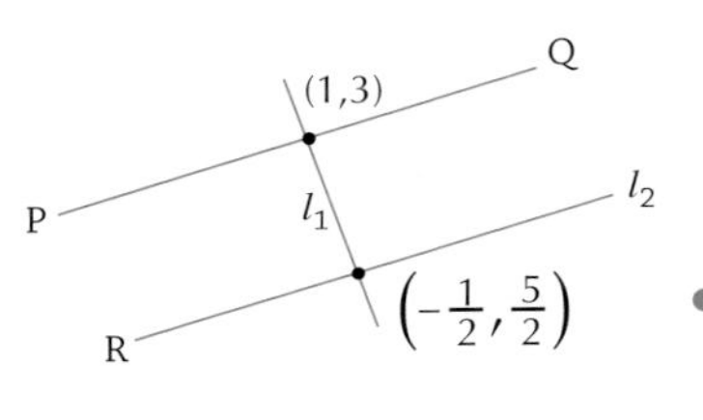

You don't have to draw one but a diagram here helps to show that the shortest (perpendicular) distance between PQ and l_2 is the length of the line segment from the midpoint of PQ (from part (a)) to the point of intersection between l_1 and l_2 (from part (c)).

$$d = \sqrt{(x_2 - x_1)^2 + (y_2 - y_1)^2}$$

$$d = \sqrt{\left(1 - \left(-\frac{1}{2}\right)\right)^2 + \left(3 - \frac{5}{2}\right)^2} \quad (\checkmark)$$

Substitute the coordinates into the 'distance' formula. You should remember this formula as it is **not** given to you in your examination. See Chapter 7 (Formulae) for more useful formulae not provided in the exam.

$$d = \sqrt{\left(\frac{3}{2}\right)^2 + \left(\frac{1}{2}\right)^2}$$

Simplify the brackets down to single fractions.

$$d = \sqrt{\frac{9}{4} + \frac{1}{4}}$$

Remember, square a fraction by squaring the numerator and squaring the denominator.

$$d = \sqrt{\frac{10}{4}} \quad (\checkmark)$$

$$= \sqrt{\frac{5}{2}}$$

Simplify the answer as far as you can using your knowledge of surds. Also acceptable here would be $\frac{\sqrt{10}}{2}$ or $\sqrt{2.5}$. See Chapter 6 (Essential skills) for more tips on working with surds.

Use properties of medians, altitudes and perpendicular bisectors

Make sure you know the difference between an altitude and a median. Altitudes are lines which go from one vertex (corner) of a triangle to the opposite side, meeting the opposite side at 90 °. To find the equation of an altitude you will usually have to find a perpendicular gradient. Medians go from a vertex of a triangle to the midpoint of the opposite side.

Example 16.3 

Triangle ABC has vertices A(–1, 12), B(–2, –5) and C(7, –2)

(a) Find the equation of the median BD. **3**

(b) Find the equation of the altitude AE. **3**

(c) Find the coordinates of the point of intersection of BD and AE. **3**

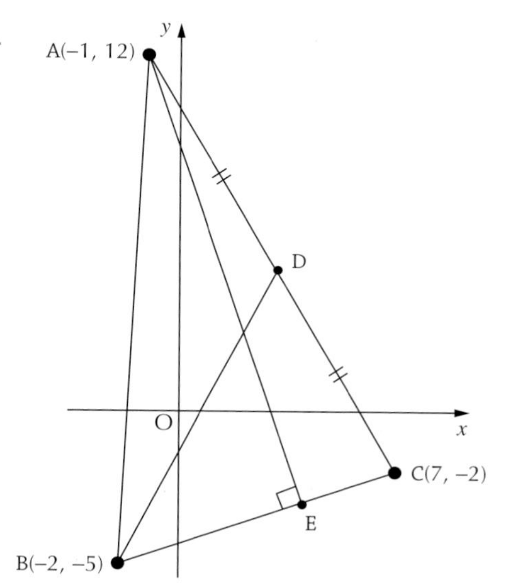

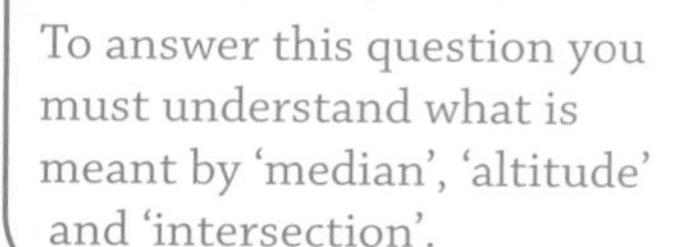

Keyword

To answer this question you must understand what is meant by 'median', 'altitude' and 'intersection'.

Hint

To find the equation of a median you need its gradient and a point on the median. We already have the coordinates of B but will need the coordinates of D before we can calculate the gradient. Since BD is a median, point D is exactly halfway between A and C.

A (a)

$$D\left(\frac{x_1 + x_2}{2}, \frac{y_1 + y_2}{2}\right)$$

$$D\left(\frac{-1+7}{2}, \frac{12+(-2)}{2}\right)$$

D (3, 5) (✓)

Find the coordinates of D using the midpoint formula. You should memorise this formula as it is not given to you in your examination. See Chapter 7 (Formulae) for more useful formulae not given to you in your exam.

$$m_{BD} = \frac{y_2 - y_1}{x_2 - x_1}$$

$$= \frac{5-(-5)}{3-(-2)}$$

$$= \frac{10}{5} = 2 \text{ (✓)}$$

Find the gradient of BD using the gradient formula. You should memorise this formula as it is **not** given to you in your examination.

$y - b = m(x - a)$

Find the equation of median BD.

$y - 5 = 2(x - 3)$

$y - 5 = 2x - 6$

$y = 2x - 1$ (✓)

Substitute the gradient of BD and one of the points, either D(3,5) or B(−2,−5) into the equation for a straight line.

It makes sense to fully simplify the equation at this point as it is often required later in the question. Other (correct) versions of the equation are also acceptable, e.g. $y - 2x = -1$ or $y - 2x + 1 = 0$.

(b)

$$m_{BC} = \frac{y_2 - y_1}{x_2 - x_1} = \frac{-2-(-5)}{7-(-2)} = \frac{3}{9} = \frac{1}{3} \ (✓)$$

$m_{AE} = -3$ since $-3 \times \frac{1}{3} = -1$ (✓)

$y - b = m(x - a)$

$y - 12 = -3\,(x - (-1))$

$y - 12 = -3\,(x + 1)$

$y - 12 = -3x - 3$

$y = -3x + 9$ (✓)

Hint

To find the equation of an altitude we need its gradient and a point on the altitude. We already have the coordinates of A but we will need to use the fact that the altitude is perpendicular to the side opposite to calculate the gradient.

Find the perpendicular gradient using $m_{\perp} = -\frac{1}{m}$ and remember to justify your answer by writing that $m \times m_{\perp} = -1$.

Find the equation of altitude AE.

Substitute the gradient of AE and the coordinates of A into the equation for a straight line.

Remove brackets and simplify. Other versions of this equation are also acceptable.

(c)

$2x - 1 = -3x + 9$ (✓)

$5x = 10$

$x = 2$ (✓)

When $x = 2$,

$y = 2\,(2) - 1 = 3$

The point of intersection is (2, 3) (✓)

Use simultaneous equations to find where two lines intersect (meet). Since we already simplified the equations in parts (a) and (b) to the form $y = \ldots$ we can set both **expressions** for y equal to each other and create an equation in x. The elimination method for simultaneous equations will also work here.

Solve to find the x-coordinate of the point of intersection.

Substitute your answer for x into either of the original straight line equations to find the corresponding y value.

Make sure you communicate clearly what you have found and give your answer as coordinates.

Use $m = \tan\theta$ to calculate a gradient or an angle

The formula $m = \tan\theta$ tells us that the gradient of a straight line is equal to the **tangent** of the angle the line makes with the *positive direction* of the x-axis, i.e. from left to right. You should memorise this important relationship as it is not given to you in your examination. The relationship can also be used in 'reverse' to find an angle, i.e. $\tan\theta = m$, which lets you find the angle using $\theta = \tan^{-1}(m)$.

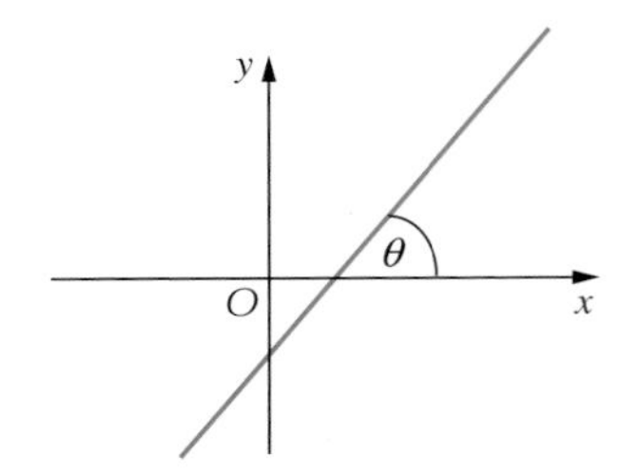

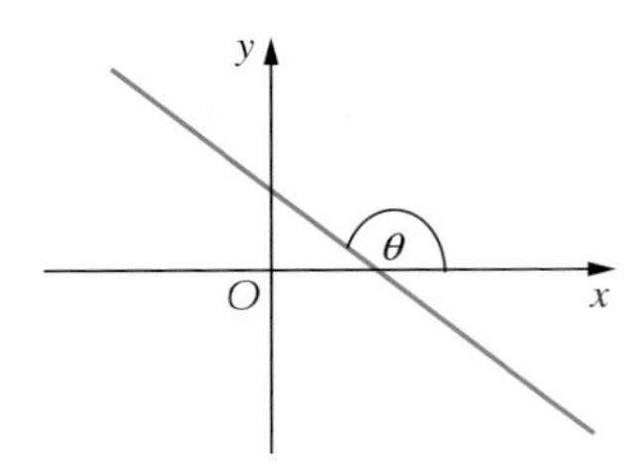

Example 16.4

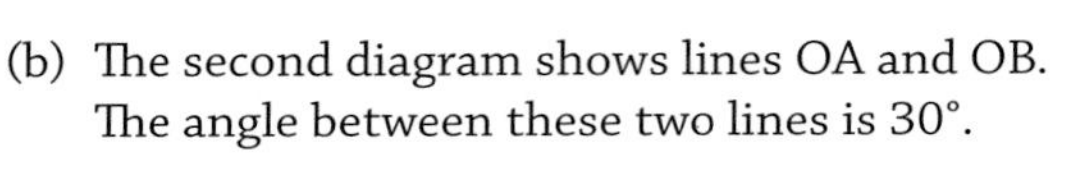

Q (a) The diagram shows line OA with equation $x - 2y = 0$

The angle between OA and the x-axis is $a°$.

Find the value of a. **3**

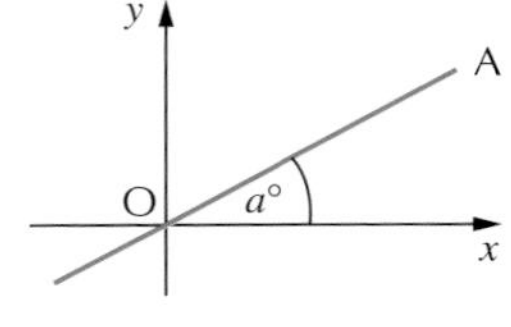

(b) The second diagram shows lines OA and OB.
The angle between these two lines is 30°.

Calculate the gradient of line OB correct to 1 decimal place. **1**

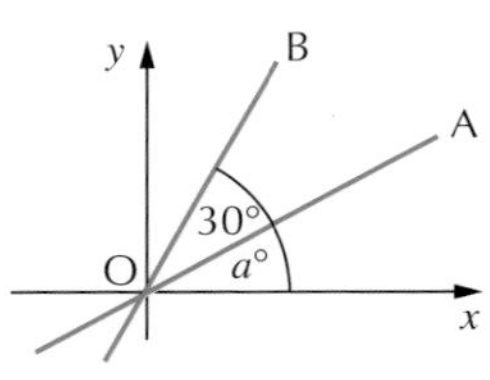

A (a)

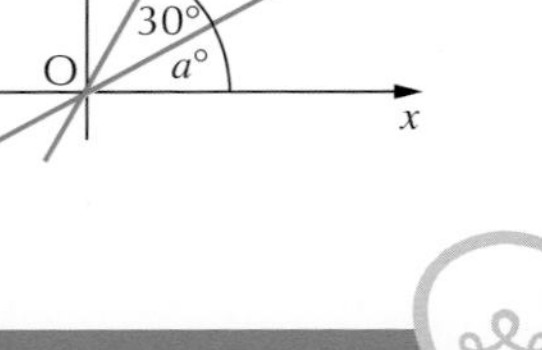

Hint

When you meet a straight line question involving an angle you should always consider $m = \tan\theta$ or $\theta = \tan^{-1}(m)$ as a possible method.

$x - 2y = 0$

$-2y = -x$

$y = \frac{1}{2}x$

Change the subject of the straight line equation to y.

$m_{OA} = \frac{1}{2}$ (✓)

When the equation is in the form $y = mx + c$, the gradient is the coefficient of x.

$\tan a° = \frac{1}{2}$ (✓)

Use $m = \tan\theta$ in 'reverse'.

$a° = \tan^{-1}\left(\frac{1}{2}\right)$

$a = 26.6°$ (✓)

(b)

$m = \tan\theta$

$m_{OB} = \tan(30 + 26.6)^\circ = \tan 56.6^\circ$

Add the angle given in the question to your answer from part (a).

$m_{OB} = 1.5$ (to 1 decimal place) (✓)

Make sure you leave your answer rounded as specified in the question.

Solve problems involving straight lines

Knowledge of common quadrilaterals is required for this type of problem. See Chapter 6 (Essential skills) for more on the properties of quadrilaterals. There are a variety of methods available to solve the final part of this question. Some require very specific working to guarantee full marks. Using a clear **vector** approach will avoid any confusion. This method is also useful in questions involving circles and other questions involving relationships between coordinates.

Example 16.5 

The diagram shows rectangle PQRS with P(7, 2) and Q(5, 6)

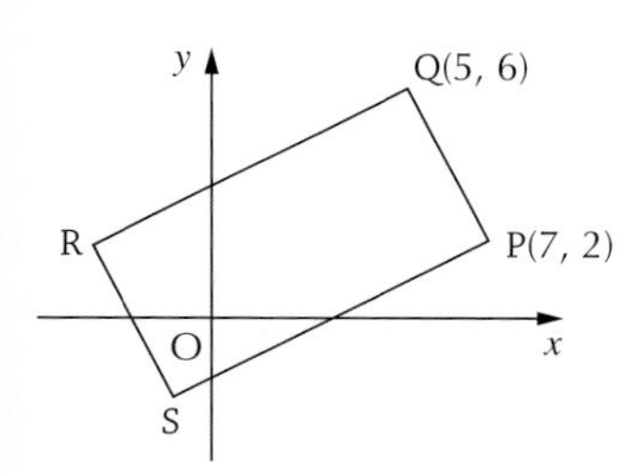

(a) Find the equation of QR **3**

(b) The line from P with the equation $x + 3y = 13$ intersects QR at T.

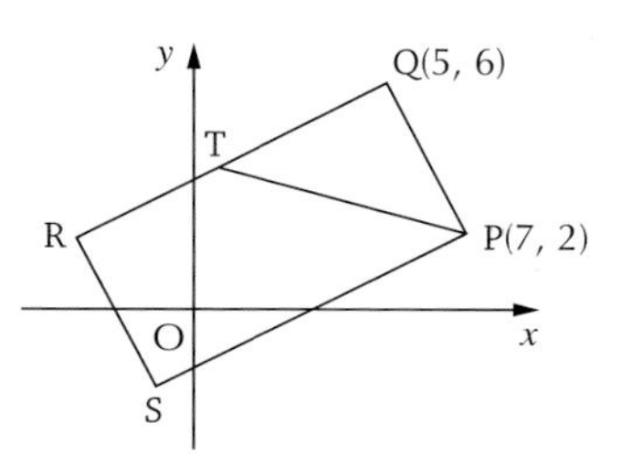

Find the coordinates of T. **3**

(c) Given that T is the midpoint of QR, find the coordinates of R and S. **3**

A (a)

$m_{PQ} = \frac{y_2 - y_1}{x_2 - x_1} = \frac{6-2}{5-7} = \frac{4}{-2} = -2$ (✓)

Since PQ and QR are adjacent sides of a rectangle, they are perpendicular. Finding m_{PQ} will let you find m_{QR}.

$m_{QR} = \frac{1}{2}$ since $\frac{1}{2} \times -2 = -1$ (✓)

Find the perpendicular gradient using $m_\perp = -\frac{1}{m}$ and remember to justify your answer by writing that $m \times m_\perp = -1$.

$y - b = m(x - a)$

$y - 6 = \frac{1}{2}(x - 5)$

$2y - 12 = x - 5$

$2y - 7 = x$ (✓)

Find the equation of QR.

Remove brackets and simplify. Other versions of this equation are also acceptable, although leaving x as the subject will make part (b) easier.

(b)

$x + 3y = 13$

$2y - 7 + 3y = 13$ (✓)

Use simultaneous equations to find where the two lines intersect. Here it is possible to replace x in the equation of PT with $2y - 7$ from part (a).

$5y = 20$

$y = 4$ (✓)

Solve to find the y-coordinate of the point of intersection.

When $y = 4$,

$x = 2(4) - 7 = 8 - 7 = 1$

Substitute your answer for y into either of the original straight line equations to find the corresponding x value.

The lines intersect at T(1, 4) (✓)

Make sure you communicate clearly what you have found and give your answer as coordinates.

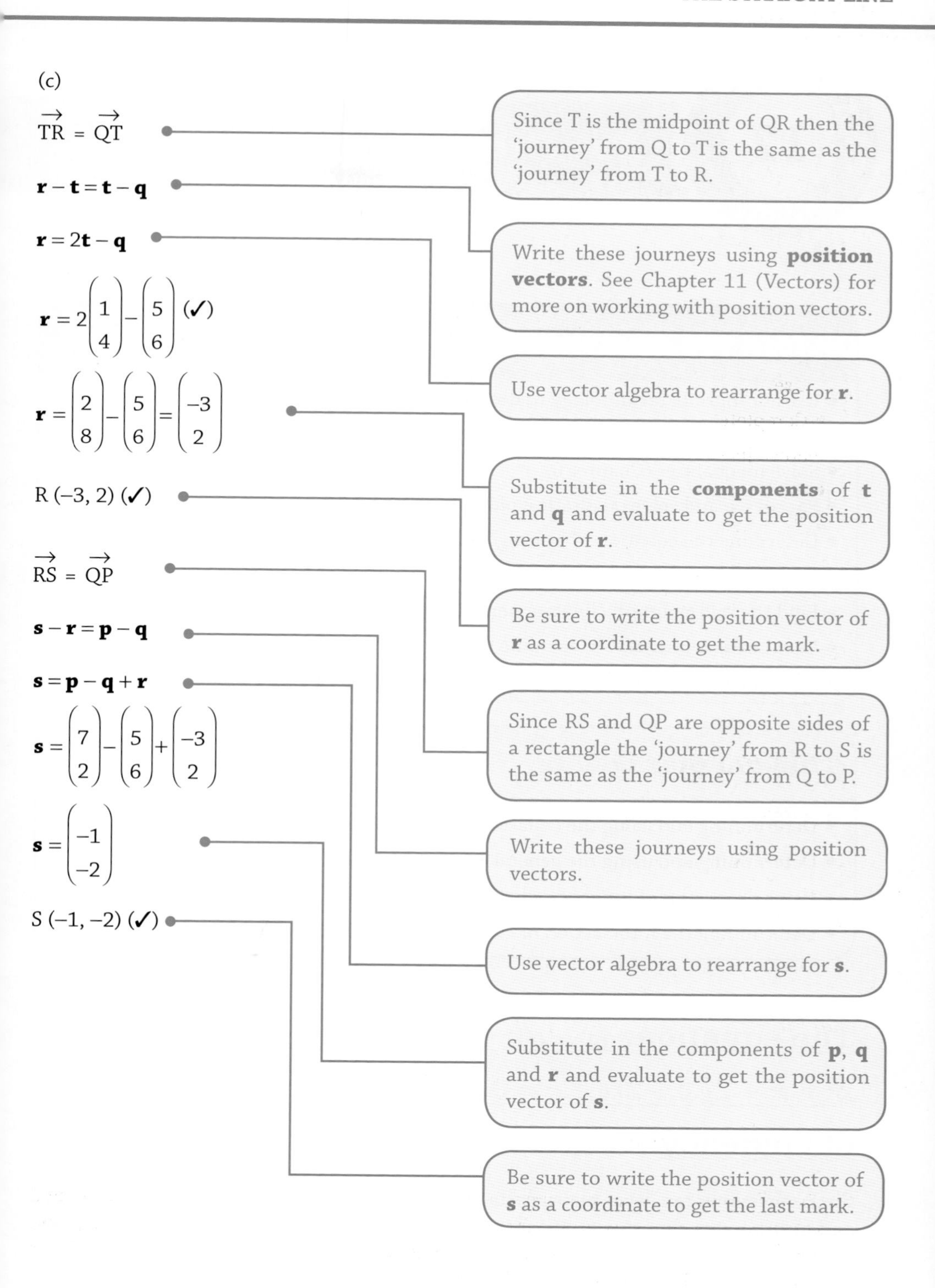

(c)

$\overrightarrow{TR} = \overrightarrow{QT}$ — Since T is the midpoint of QR then the 'journey' from Q to T is the same as the 'journey' from T to R.

$\mathbf{r} - \mathbf{t} = \mathbf{t} - \mathbf{q}$ — Write these journeys using **position vectors**. See Chapter 11 (Vectors) for more on working with position vectors.

$\mathbf{r} = 2\mathbf{t} - \mathbf{q}$ — Use vector algebra to rearrange for **r**.

$\mathbf{r} = 2\begin{pmatrix}1\\4\end{pmatrix} - \begin{pmatrix}5\\6\end{pmatrix}$ (✓)

$\mathbf{r} = \begin{pmatrix}2\\8\end{pmatrix} - \begin{pmatrix}5\\6\end{pmatrix} = \begin{pmatrix}-3\\2\end{pmatrix}$ — Substitute in the **components** of **t** and **q** and evaluate to get the position vector of **r**.

R (–3, 2) (✓) — Be sure to write the position vector of **r** as a coordinate to get the mark.

$\overrightarrow{RS} = \overrightarrow{QP}$ — Since RS and QP are opposite sides of a rectangle the 'journey' from R to S is the same as the 'journey' from Q to P.

$\mathbf{s} - \mathbf{r} = \mathbf{p} - \mathbf{q}$ — Write these journeys using position vectors.

$\mathbf{s} = \mathbf{p} - \mathbf{q} + \mathbf{r}$ — Use vector algebra to rearrange for **s**.

$\mathbf{s} = \begin{pmatrix}7\\2\end{pmatrix} - \begin{pmatrix}5\\6\end{pmatrix} + \begin{pmatrix}-3\\2\end{pmatrix}$

$\mathbf{s} = \begin{pmatrix}-1\\-2\end{pmatrix}$ — Substitute in the components of **p**, **q** and **r** and evaluate to get the position vector of **s**.

S (–1, –2) (✓) — Be sure to write the position vector of **s** as a coordinate to get the last mark.

The circle

Before you start this chapter you should be able to:

- Use appropriate terminology to describe a circle, i.e. **centre**, radius, diameter
- Complete the square in a quadratic **expression** (Chapter 10)
- Solve inequalities (Chapter 6)
- **Factorise** and solve quadratic equations (Chapter 6)
- Determine the equation of a straight line (Chapter 16)
- Use simultaneous equations to find where two lines **intersect** (Chapter 16)
- Find the mid-point of a line segment (Chapter 16)
- Use the **discriminant** to examine the **roots** of a quadratic equation (Chapter 12)
- Use the distance formula to calculate the distance between two points.

This chapter covers:

- Determining and using the equation of a circle $(x - a)^2 + (y - b)^2 = r^2$
- Determining and using the general equation of a circle $x^2 + y^2 + 2gx + 2fy + c = 0$
- Using properties of tangency when solving problems
- Determining the intersection of circles or a line and a circle.

The equations $(x - a)^2 + (y - b)^2 = r^2$ and $x^2 + y^2 + 2gx + 2fy + c = 0$ and the properties of tangency

Circle questions can have many parts. The next example is worth 12 marks in total and incorporates almost everything in the circle topic. Take each part step-by-step and remember that in many cases you might still be able to attempt later parts even if you have not completed one of the earlier parts.

Example 17.1 

A circle C_1 has equation $x^2 + y^2 + 2x + 4y - 27 = 0$.

(a) Write down the centre and calculate the radius of C_1. **2**

(b) The point P(3, 2) lies on the circle C_1.

Find the equation of the **tangent** at P. **3**

(c) A second circle C_2 has centre (10, −1). The radius of C_2 is half of the radius of C_1.

Show that the equation of C_2 is $x^2 + y^2 - 20x + 2y + 93 = 0$. **3**

In any circle question where you are given the equation in the form $x^2 + y^2 + 2gx + 2fy + c = 0$ there is a good chance you are going to need the centre and radius. All you have to do is work these out to get the first two marks.

(d) Show that the tangent found in part (*b*) is also a tangent to circle C_2. **4**

(a)

$x^2 + y^2 + 2x + 4y - 27 = 0$

$2g = 2 \quad 2f = 4$

When a circle equation is in the form $x^2 + y^2 + 2gx + 2fy + c = 0$, $2g$ is the **coefficient** of x and $2f$ is the coefficient of y.

$g = 1 \quad f = 2$

$C_1(-1, -2)$ (✓)

The centre is the point $(-g, -f)$. This result is on the formula sheet given to you in your examination.

$r = \sqrt{g^2 + f^2 - c}$

This formula is also given to you in your examination. Use it to calculate the radius of a circle when the equation is given in the form $x^2 + y^2 + 2gx + 2fy + c = 0$.

$r_{C_1} = \sqrt{1^2 + 2^2 - (-27)}$

Avoid losing marks here; the signs matter so make sure you use g and f. Take care when c is negative.

$r_{C_1} = \sqrt{32}$

$= 4\sqrt{2}$ (✓)

It is always a good idea to simplify surds. In this question the radius r_{C_1} is needed further on.

(b)

Hint

A radius can be drawn from the centre of a circle to any point on the circle. This radius is perpendicular to the tangent at that point. Although not always given in the question, a diagram like the one below can help you visualise the situation.

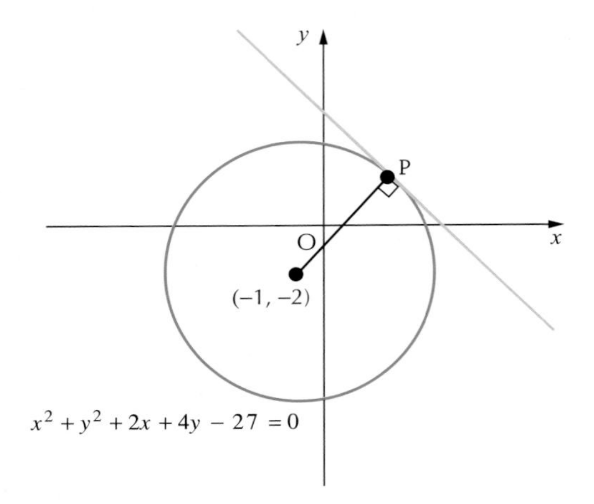

$C_1\ (-1, -2)\quad P(3, 2)$

$m_{\textbf{radius}} = \dfrac{y_2 - y_1}{x_2 - x_1} = \dfrac{2-(-2)}{3-(-1)} = 1\ (\checkmark)$

$m_{\textbf{tangent}} = -1$ since $1 \times -1 = -1\ (\checkmark)$

Find the perpendicular **gradient** using $m_{\perp} = -\frac{1}{m}$ and remember to justify your answer properly, i.e. making sure you explain why you know the lines are perpendicular. You should show the numerical value of each gradient. If you write any less than what is shown here you risk losing a mark.

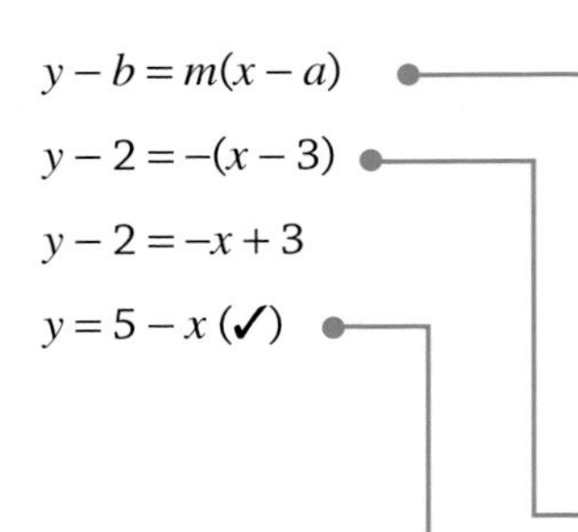

$y - b = m(x - a)$

$y - 2 = -(x - 3)$

$y - 2 = -x + 3$

$y = 5 - x\ (\checkmark)$

Find the equation of the tangent at P.

Make sure it is the coordinates of P(3,2) you substitute in and not the coordinates of the centre.

Remove brackets and simplify. Other versions of this equation are also acceptable.

(c)

$$r_{C_2} = \frac{1}{2} r_{C_1}$$

$$r_{C_2} = \frac{1}{2} \times 4\sqrt{2} = 2\sqrt{2} \ (\checkmark)$$

$$(x - a)^2 + (y - b)^2 = r^2$$

$$(x - 10)^2 + (y - (-1))^2 = (2\sqrt{2})^2 \ (\checkmark)$$

$$(x - 10)^2 + (y + 1)^2 = 8$$

$$(x - 10)(x - 10) + (y + 1)(y + 1) = 8$$

$$x^2 - 10x - 10x + 100 + y^2 + y + y + 1 = 8$$

$x^2 + y^2 - 20x + 2y + 93 = 0$ as required (✓)

When you know the centre (a, b) and radius of a circle you should substitute them into this equation, which is given to you in your examination. Use this to find the equation of circle C_2.

Always take a line of working to substitute in the values. Take care with the radius, which, in this example, must go in brackets.

The right-hand side becomes 8 since $(2\sqrt{2})^2 = 2^2 \times \sqrt{2}^2 = 4 \times 2 = 8$.

Take as many steps as you need to expand and simplify the brackets.

Carefully follow through the algebra until your equation looks like the one you were asked to 'show'.

Hint

With all 'show that' questions you can't use the given result in your proof. You must start somewhere else, usually with another equation, and work towards the given result.

(d)

Hint

To show that a line is a tangent to a circle you need to prove that there is only one point of contact between the line and the circle. This can be done using the discriminant ($b^2 - 4ac$). See Chapter 12 (Polynomials) for more on working with the discriminant.

There are three possibilities:

Description	Condition on discriminant	Visual representation
The line is a tangent to the circle	$b^2 - 4ac = 0$	
The line is a **chord**	$b^2 - 4ac > 0$	
The line doesn't touch or cross the circle	$b^2 - 4ac < 0$	

In this question it is possible to get all the marks for part (d), even if you haven't completed part (c) since you are given the equation of C_2.

$x^2 + y^2 - 20x + 2y + 93 = 0$

$y = 5 - x$

$x^2 + (5 - x)^2 - 20x + 2(5 - x) + 93 = 0$ (✓)

$x^2 + (5 - x)(5 - x) - 20x + 10 - 2x + 93 = 0$

$x^2 + 25 - 5x - 5x + x^2 - 20x + 10 - 2x + 93 = 0$

$2x^2 - 32x + 128 = 0$ (✓)

$x^2 - 16x + 64 = 0$

$a = 1$

$b = -16$

$c = 64$

$b^2 - 4ac$

$= (-16)^2 - 4(1)(64)$ (✓)

$= 256 - 256$

$= 0$

$b^2 - 4ac = 0$ therefore the line is a tangent. (✓)

Clearly communicate that you know a discriminant of zero means the line is a tangent.

Rearrange the **linear** equation from part (b) to make either x or y the subject and substitute into the circle equation. Here we have substituted for y.

Take as many lines of working as you need to remove the brackets and simplify.

This will probably be the longest single line of working you write in your examination. Take care and don't miss out any terms.

Write the equation in standard quadratic form taking care not to miss out the '= 0'.

Simplify by dividing throughout by the highest common **factor**. This won't affect your result but will make the numbers a little bit easier to work with.

a and b are the coefficients of x^2 and x. c is the **constant** (number) term. Make sure you include the correct sign for each.

Substitute a, b and c into the discriminant. Make sure any negative b values go in brackets, -16^2 is **not** the same as $(-16)^2$.

This is a non-calculator question so you will have to work out $(-16)^2$ any way you can. Long multiplication is a good idea.

Determine the intersection of circles or a line and a circle

To determine whether or not two circles intersect we compare the distance between the two centres with the sum of the radii. There are five possibilities:

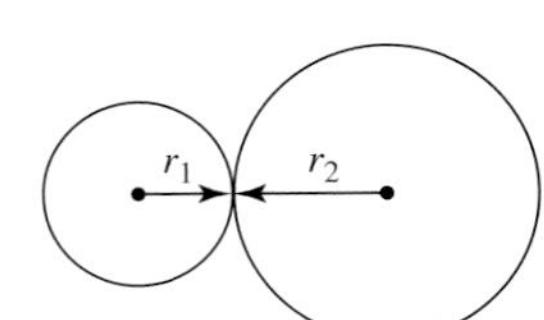

1 The circles meet externally at one point.

Here the distance between the centres equals the sum of the radii, i.e. $d = r_1 + r_2$.

2 It is possible that the circles meet at only one point and that one circle is positioned inside the other. If this is the case then the distance between the centres will equal the *difference* between the radii, i.e. $d = r_1 - r_2$.

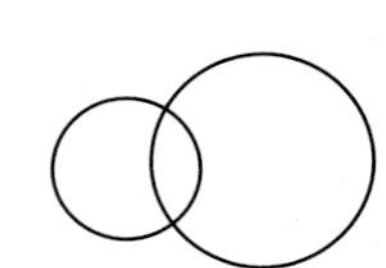

3 The circles intersect at two points.

In this situation the distance between the centres is less than the sum of the radii, i.e. $d < r_1 + r_2$.

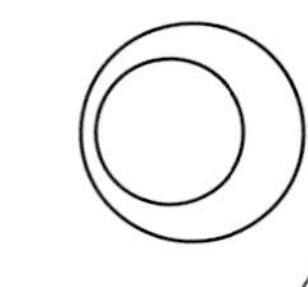

4 The circles do not touch.

In this situation the distance between the centres is greater than the sum of the radii, i.e. $d > r_1 + r_2$.

5 The circles do not touch and one circle is contained in the other. In this case the distance between the centres is less than the *difference* between the radii, i.e. $d < r_1 - r_2$.

Example 17.2 

Q (a) Relative to a suitable set of coordinate axes, Diagram 1 shows the line $2x - y + 5 = 0$ intersecting the circle $x^2 + y^2 - 6x - 2y - 30 = 0$ at the points P and Q.

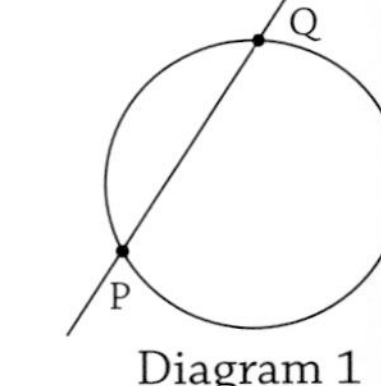

Diagram 1

Find the coordinates of P and Q. **6**

(b) Diagram 2 shows the circle from (*a*) and a second **congruent** circle, which also passes through P and Q.

Determine the equation of this second circle. **6**

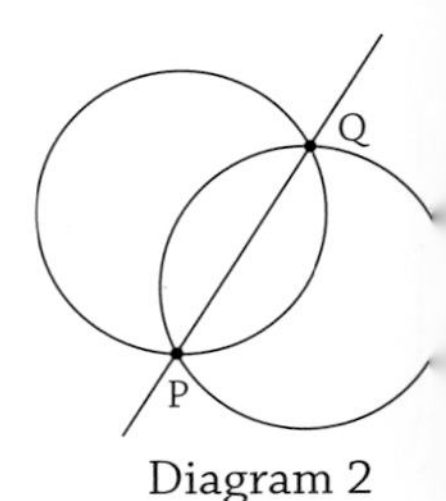

Diagram 2

Hint

To find the points of contact between a line and a circle we substitute the linear equation into the circle equation (rearranging first if necessary) → simplify the result into standard quadratic form → factorise and solve to find the x-coordinates of the points of intersection → substitute the x values into the linear equation to get the corresponding y values.

Keyword

If two shapes are '**congruent**' then they are exactly the same shape and size.

A (a)

$2x - y + 5 = 0$

$2x + 5 = y$ (✓)

Rearrange the linear equation so that y is the subject.

$x^2 + y^2 - 6x - 2y - 30 = 0$

$x^2 + (2x + 5)^2 - 6x - 2(2x + 5) - 30 = 0$ (✓)

Replace y in the circle equation with $2x + 5$ from the linear equation.

$x^2 + (2x + 5)(2x + 5) - 6x - 2(2x + 5) - 30 = 0$

$x^2 + 4x^2 + 10x + 10x + 25 - 6x - 4x - 10 - 30 = 0$

$5x^2 + 10x - 15 = 0$ (✓)

Carefully simplify and leave in standard quadratic form. To get full marks you must include the '= 0'.

$x^2 + 2x - 3 = 0$

Simplify by dividing throughout by the highest common factor. This won't affect your solutions but will make the numbers a little bit easier to work with.

$(x + 3)(x - 1) = 0$ (✓)

Factorise, remembering to check that multiplying out gets you back to the original equation.

$x + 3 = 0$ and $x - 1 = 0$

$x = -3$ and $x = 1$ (✓)

Set each factor equal to zero and solve for x.

When $x = -3$, $y = 2(-3) + 5 = -1$

When $x = 1$, $y = 2(1) + 5 = 7$ (✓)

Substitute your x values into the linear equation to find the corresponding y values.

The line and circle intersect at P (–3, –1) and Q (1, 7)

Give your answers as coordinates. Looking at the coordinates we can tell P and Q apart from the diagram in the question since point Q is higher up and further right than point P.

(b)

$x^2 + y^2 - 6x - 2y - 30 = 0$

$2g = -6 \qquad 2f = -2$

$g = -3 \qquad f = -1$

$C_1\,(3, 1)$ (✓)

This is the centre of the original circle.

Hint

To find the equation of a circle you should substitute the centre and radius into the equation $(x - a)^2 + (y - b)^2 = r^2$.

$r = \sqrt{g^2 + f^2 - c}$

$r = \sqrt{(-3)^2 + (-1)^2 - (-30)} = \sqrt{40}$ (✓)

There is no need to simplify this surd here since the radius will go into the equation $(x - a)^2 + (y - b)^2 = r^2$ where it will become $\sqrt{40^2} = 40$.

$M\left(\dfrac{-3+1}{2}, \dfrac{-1+7}{2}\right)$

$M\,(-1, 3)$ (✓)

Since the circles are congruent the line PQ is a line of symmetry. The midpoint of PQ is also the midpoint between the centres of the circles.

$C_2(x, y)$

$M(-1, 3)$

$C_1(3, 1)$

$\overrightarrow{MC} = \overrightarrow{C_1M}$

The 'journey' from the midpoint of PQ to the centre of the second circle is the same as the 'journey' from the centre of the first circle to the midpoint of PQ.

$\mathbf{c}_2 - \mathbf{m} = \mathbf{m} - \mathbf{c}_1$

Write these 'journeys' using **position vectors**.

$\mathbf{c}_2 = 2\mathbf{m} - \mathbf{c}_1$

Use vector algebra to rearrange for $\mathbf{c}_2$.

$\mathbf{c}_2 = 2\begin{pmatrix}-1\\3\end{pmatrix} - \begin{pmatrix}3\\1\end{pmatrix}$ (✓)

$\mathbf{c}_2 = \begin{pmatrix}-2\\6\end{pmatrix} - \begin{pmatrix}3\\1\end{pmatrix} = \begin{pmatrix}-5\\5\end{pmatrix}$

Substitute in the **components** of **m** and $\mathbf{c}_1$ and evaluate to get the position vector of $\mathbf{c}_2$.

$C_2\,(-5, 5)$ (✓)

To get this mark you must clearly state the coordinates of the centre of the second circle.

$(x - a)^2 + (y - b)^2 = r^2$

Find the equation of the second circle.

$(x - (-5))^2 + (y - 5)^2 = (\sqrt{40})^2$

$(x + 5)^2 + (y - 5)^2 = 40$ (✓)

You must fully simplify the radius to get the final mark.

Example 17.3

Q

(a) Find P and Q, the points of intersection of the line $y = 3x - 5$ and the circle C_1 with equation $x^2 + y^2 + 2x - 4y - 15 = 0$. **4**

(b) T is the centre of C_1. **3**

Show that PT and QT are perpendicular.

(c) A second circle C_2 passes through P, Q and T.

Find the equation of C_2. **3**

Keyword

You must understand the words 'intersection' and 'perpendicular' to be able to answer this question correctly.

A (a)

$x^2 + y^2 + 2x - 4y - 15 = 0, \quad y = 3x - 5$

$x^2 + (3x - 5)^2 + 2x - 4(3x - 5) - 15 = 0$ (✓)

> Substitute the linear equation into the circle equation. Remember, sometimes you will have to change the subject of the linear equation to either y or x before you can substitute.

$x^2 + (3x - 5)(3x - 5) + 2x - 4(3x - 5) - 15 = 0$

$x^2 + 9x^2 - 15x - 15x + 25 + 2x - 12x + 20 - 15 = 0$

> Take as many lines of working as you need to expand and simplify the equation. Mistakes here will make the later parts of this question much more difficult.

$10x^2 - 40x + 30 = 0$ (✓)

> Write in standard quadratic form and don't forget the '= 0' or you will lose a mark.

$x^2 - 4x + 3 = 0$

> Dividing throughout by 10 won't affect your answer and will make the equation easier to factorise.

$(x - 3)(x - 1) = 0$

> Always check you have factorised correctly by multiplying and making sure you get the quadratic you started with.

$x-3=0$ and $x-1=0$

$x=3$ and $x=1$ (✓)

Set each factor equal to zero and solve to find the x-coordinates of the points of intersection.

When $x=3$, $y=3(3)-5=4$

When $x=1$, $y=3(1)-5=-2$

Substitute your x values into the linear equation to find the corresponding y values.

The line and circle intersect at P (3, 4) and Q (1, –2) (✓)

Give your answers as coordinates. In this question it doesn't matter which point you call P and which you call Q.

(b)

$x^2+y^2+2x-4y-15=0$

$2g=2$ and $2f=-4$

$g=1$ and $f=-2$

T (–1, 2) (✓)

Hint

To show that PT and QT are perpendicular we need to show that their gradients multiply to give –1. We have P and Q from part (a), now we need to find T, the centre of C_1.

When a circle equation is in the form $x^2+y^2+2gx+2fy+c=0$, $2g$ is the coefficient of x and $2f$ is the coefficient of y.

The centre is the point $(-g, -f)$. This result is on the formula sheet given to you in your examination. Use this to find the centre of C_1.

P(3, 4) T(–1, 2)

$$m_{PT}=\frac{y_2-y_1}{x_2-x_1}=\frac{4-2}{3-(-1)}=\frac{2}{4}=\frac{1}{2}$$

Q(1, –2) T(–1, 2)

$$m_{QT}=\frac{y_2-y_1}{x_2-x_1}=\frac{-2-2}{1-(-1)}=\frac{-4}{2}=-2 \quad (✓)$$

Carefully find the gradients of PT and QT and label them m_{PT} and m_{QT} to avoid mixing them up later on.

$m_{PT}\times m_{QT}=\frac{1}{2}\times -2=-1$ therefore

PT is perpendicular to QT (✓)

Communicate your answer properly, i.e. making sure you explain why you know the lines are perpendicular. You should show the numerical value of each gradient. If you write any less than what is shown here you risk losing a mark.

(c)

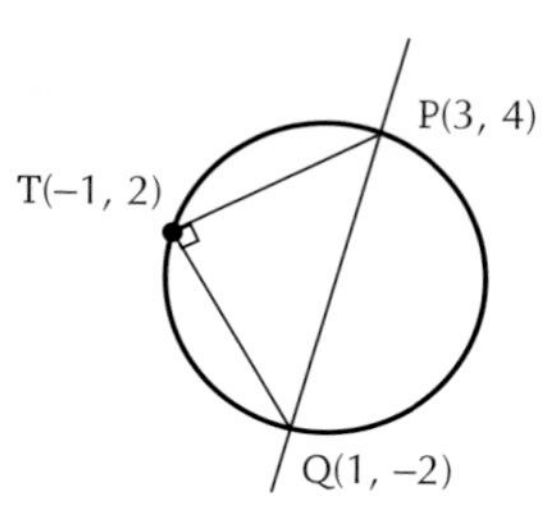

Hint

A sketch can be very helpful in questions like this and will let you see connections you might otherwise miss.

Since PT and QT are perpendicular, i.e. they meet at 90°, your knowledge of the angles in a circle tells you that PQ is a diameter of the new circle.

$$C_2\left(\frac{3+1}{2}, \frac{4+(-2)}{2}\right)$$

$C_2\ (2, 1)$ (✓)

The centre of the new circle is the mid-point of the line joining points P and Q.

$$r_{C_2} = \sqrt{(3-2)^2 + (4-1)^2} = \sqrt{1^2 + 3^2} = \sqrt{10} \ (✓)$$

Use the distance formula to calculate radius, i.e. the distance between $C_2\ (2, 1)$ and either P or Q.

$C_2\ (2, 1) \qquad r_{C_2} = \sqrt{10}$

$$(x-a)^2 + (y-b)^2 = r^2$$

$$(x-2)^2 + (y-1)^2 = \sqrt{10}^2$$

Find the equation of circle C_2.

$$(x-2)^2 + (y-1)^2 = 10 \ (✓)$$

You must fully simplify the radius to get the final mark.

Sequences

Before you start this chapter you should be able to:

- Recognise commonly used **sequences** like the square and cube numbers.

This chapter covers:

- The terminology and notation associated with sequences
- Using and determining n^{th} term formulae
- Determining a **recurrence relation** from given information
- Using a recurrence relation to calculate a required term
- Finding and interpreting a **limit** of a sequence, where it exists.

Working with sequences, recurrence relations and their limits

A sequence is an ordered list of numbers generated by a set rule. Each number in the sequence is called a term. The set rule must show a relationship between any given term and the term following it. Sequences like (2, 4, 6, ..., 70) are called finite sequences. The ellipsis (...) in the sequence followed by the final number indicates that some of the terms have been missed out.
Other sequences are infinite where the terms go on forever. (3, 6, 9, 12, ...) is an example of an infinite sequence, the three dots with nothing after them indicate the sequence never ends.
Terms are often denoted by $u_1, u_2, u_3, \ldots, u_{n-1}, u_n, u_{n+1}$, where u_1 is the first term, u_2 is the second term and u_n is the n th term. Where appropriate, the final term is denoted by l. In many sequences u_0 will be defined. This is the notation for an initial (or starting) value. Usually this value is not generated by the sequence, meaning that u_1 will be the first term actually generated by the sequence.

Example 18.1 

A sequence is defined by $u_{n+1} = -\frac{1}{2}u_n$ with $u_0 = -16$.

(a) Determine the values of u_1 and u_2. **1**

(b) A second sequence is given by 4, 5, 7, 11, ...
It is generated by the recurrence relation $v_{n+1} = pv_n + q$ with $v_1 = 4$.
Find the values of p and q. **3**

(c) Either the sequence in (a) or the sequence in (b) has a limit.

(i) Calculate this limit.
(ii) Why does this other sequence not have a limit? **3**

(a)

$u_{n+1} = -\frac{1}{2}u_n$ with $u_0 = -16$

$u_1 = -\frac{1}{2}u_0 = -\frac{1}{2}(-16) = 8$

Substitute the value of u_0 into the recurrence relation to generate u_1.

$u_2 = -\frac{1}{2}u_1 = -\frac{1}{2}(8) = -4$ (✓)

Repeat, this time substituting your value for u_1 to generate u_2.

Hint

Simultaneous equations are often useful when you are trying to find two unknowns. In this example you are required to use the recurrence relation, along with the first few terms of the sequence, to 'build' the equations.

(b)

$v_{n+1} = pv_n + q$

$5 = 4p + q$

Substitute the first two terms of the sequence, i.e. 4 and 5, into the recurrence relation in place of v_n and v_{n+1}.

$7 = 5p + q$ (✓)

Substitute terms two and three of the sequence, i.e. 5 and 7, into the recurrence relation in place of v_n and v_{n+1}.

$p = 2$ (✓)

Subtracting the first equation from the second one eliminates q and lets you find p.

When $p = 2$, $q = 5 - 4p = 5 - 4(2) = -3$ (✓)

Substitute the value of p into either of the two original equations to find q.

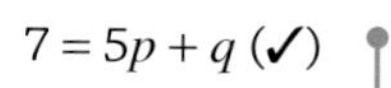

$v_{n+1} = 2v_n - 3$

(c) (i)

Hint

A recurrence relation of the form $u_{n+1} = au_n + b$ will have a limit if $-1 < a < 1$. This limit can be found using the formula $L = \frac{b}{1-a}$, which you should memorise. See Chapter 7 (Formulae) for more formulae not given to you in your exam.

$u_{n+1} = -\frac{1}{2}u_n$ has a limit since $-1 < -\frac{1}{2} < 1$

Get into the habit of justifying when a limit exists. Be careful, writing $-1 < a < 1$ will not get a mark unless you have defined a somewhere in your working.

$$L = \frac{b}{1-a} = \frac{0}{1-\left(-\frac{1}{2}\right)} = 0 \quad (✓)(✓)$$

Be prepared to evaluate limits similar to this one without using your calculator.

(ii)

$v_{n+1} = 2v_n - 3$ does not have a limit since $2 > 1$ (✓)

Think carefully about how to explain why this sequence doesn't have a limit.

Applying calculus skills to optimisation and area

Before you start this chapter you should be able to:

- Differentiate algebraic, trigonometric and simple **composite functions** (Chapter 14)
- Find the **stationary points** of a curve (Chapter 14)
- **Integrate** algebraic, trigonometric and simple composite **functions** (Chapter 15).

This chapter covers:

- Finding the greatest/least **values** of a function on a closed interval
- Determining the optimal solution for a given problem (**optimisation**)
- Solving problems using rates of change
- Finding the area between a curve and the x-axis
- Finding the area between a straight line and a curve or between two curves.

Finding the greatest/least values of a function on a closed interval

This type of question combines skills from across the course. You could be asked to **factorise** and solve a polynomial then go on to either find the maximum/minimum values, or even to sketch the graph of the polynomial. See Chapter 12 (Polynomials) for more on factorising and solving polynomials and Chapter 14 (**Differentiation**) for curve sketching.

Example 19.1 

A function f is defined by the formula $f(x) = 2x^3 - 7x^2 + 9$ where x is a real number.

(a) Show that $(x - 3)$ is a **factor** of $f(x)$, and hence factorise $f(x)$ fully. **5**

(b) Find the coordinates of the points where the curve with equation $y = f(x)$ crosses the x- and y-axes. **2**

Find the greatest and least values of f in the interval $-2 \leq x \leq 2$. **5**

A (a)

Keyword

The words 'greatest' and 'least' suggest that you should find **turning points** and that you should find the value of the function at both ends of the given interval.

$f(x) = 2x^3 - 7x^2 + 9$

	x^3	x^2	x	c	
3	2	−7	0	9	(✓)
		6	−3	−9	
	2	−1	−3	0	(✓)

$x - 3$ is a factor as the remainder is zero. (✓)

$f(x) = (x - 3)(2x^3 - x - 3)$ (✓)

$f(x) = (x - 3)(2x - 3)(x + 1)$ (✓)

Decreasing powers of x. Here c represents the **constant** (number) term.

Coefficients of x^3, x^2, x and the constant. If a term is missing you *must* include a zero, in this case the x is missing.

Add vertically then multiply diagonally from left to right.

The number in the box is the remainder when the polynomial is divided by $(x - 3)$. A zero here means $(x - 3)$ is a factor.

Communication is essential here to get the mark.

Use the numbers from the bottom row of the synthetic division table as the coefficients of the quadratic factor, $2x^2 - x - 3$.

Factorise the quadratic factor to leave the function in fully factorised form.

(b)

$f(x) = 2x^3 - 7x^2 + 9$

Crosses y-axis when $x = 0$

$f(0) = 9$

Substitute $x = 0$ into the function to find where the graph of the function crosses the y-axis.

Curve crosses the y-axis at $(0, 9)$ (✓)

Crosses x-axis when $y = 0$

$(x - 3)(2x - 3)(x + 1) = 0$

Substitute $y = 0$ into the factorised version function to find where the graph of the function crosses the x-axis.

$x - 3 = 0$ and $2x - 3 = 0$ and $x + 1 = 0$

$x = 3$ and $x = \frac{3}{2}$ and $x = -1$

Set each factor equal to zero and solve to find where the curve crosses the x-axis.

Curve crosses the x-axis at $(-1, 0)$ $\left(\frac{3}{2}, 0\right)$ $(3, 0)$(✓)

(c)

Hint

The greatest and least values might occur at the maximum and minimum turning points. However, this is not always the case. You must also find the value of the function at the endpoints of the given interval.

$f(x) = 2x^3 - 7x^2 + 9, \qquad -2 \leq x \leq 2$

$f'(x) = 6x^2 - 14x$

Differentiate to get a formula for the **gradient** of the function.

$6x^2 - 14x = 0$ (✓)

Set the **derivative** equal to zero.

$2x(3x - 7) = 0$

$2x = 0$ and $3x - 7 = 0$

$x = 0$ and $x = \frac{7}{3}$ (✓)

Factorise and solve to find the x-coordinates of the stationary points. Since $\frac{7}{3} > 2$ the second stationary point is outside the interval given in the question and the turning point at $x = \frac{7}{3}$ can be ignored.

$f(0) = 9$

We already worked out $f(0) = 9$ in part (a). This is where the curve crosses the y-axis.

$f(-2) = 2(-2)^3 - 7(-2)^2 + 9 = -35$ (✓)

$f(2) = 2(2)^3 - 7(2)^2 + 9 = -3$

Find the value of the function at *both* of the end points by substituting into $f(x)$.

Greatest value = 9 (✓)

Least value = – 35 (✓)

This corresponds to the maximum turning point of $f(x)$.

This sketch of $f(x)$ shows that, although the function has a minimum turning point when $x = \frac{7}{3}$, it is outwith the interval we are working in. Plus, you can see that the value of the function is much lower when $x = -2$.

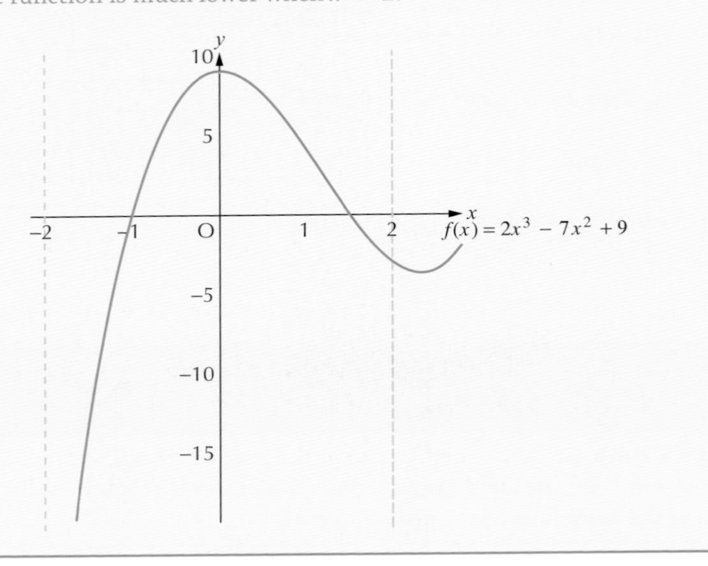

Solving problems using rates of change

Differentiation is used to help us understand and predict the effects on one variable as another variable changes. Often one of these variables is time, for example, we might investigate how distance changes with time, or how speed changes with time, although there are many other relationships that can be understood using differentiation. When we talk about how one variable changes with another we use the phrase '**rate of change**'. When you see this phrase it is very likely that you will be expected to differentiate.

Example 19.2 

Q The volume of a sphere is given by the formula $V = \frac{4}{3}\pi r^3$.

Find the **exact value** of the rate of change of V with respect to r, at $r = 2$. **2**

Keyword

'Rate of change' tells you that you are going to have to differentiate.

Hint

Don't be put off if π appears in a function, it doesn't affect how you differentiate and you should just treat it as another number.

$V = \frac{4}{3}\pi r^3$

$\frac{dV}{dr} = 4\pi r^2$ (✓) — Differentiate, making sure you use the variables given in the question, i.e. V and r.

When $r = 2$, $\frac{dV}{dr} = 4\pi(2)^2 = 16\pi$ (✓) — Substitute $r = 2$ into the derivative and evaluate. Leaving your answer in terms of π means it is exact.

Finding the area between a curve and the x-axis

Integration allows us to find areas of shapes that are not made up of simple polygons or parts of a circle. Be careful when finding areas that are part above and part below the x-axis, as areas below the x-axis are negative. In these situations you should split the shape up and integrate it a section at a time.

Example 19.3 

The graph shown has equation $y = x^3 - 6x^2 + 4x + 1$. The total shaded area is bounded by the curve, the x-axis, the y-axis and the line $x = 2$.

(*a*) Calculate the shaded area labelled S. **4**

(*b*) Hence find the total shaded area. **3**

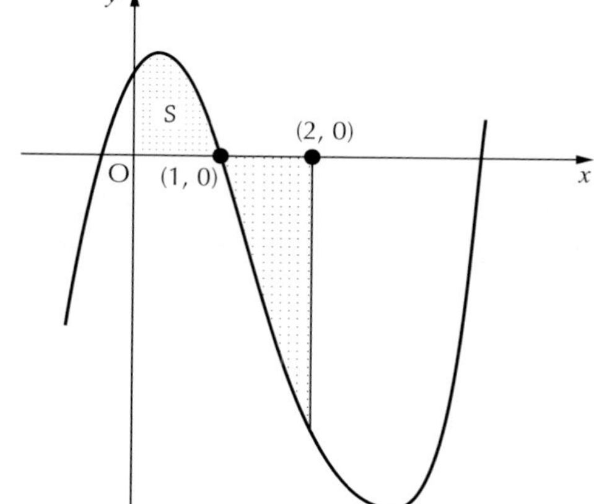

A (a)

$y = x^3 - 6x^2 + 4x + 1$

$$\int_0^1 (x^3 - 6x^2 + 4x + 1)dx \quad (✓)$$

Prepare to integrate the function between the **limits** of $x = 0$ and $x = 1$. This will give the shaded area S. To be sure of full marks you must include the dx.

$$= \left[\frac{x^4}{4} - \frac{6x^3}{3} + \frac{4x^2}{2} + x\right]_0^1 \quad (✓)$$

Integrate by 'adding one to the **index** then dividing by the new index'. Remember that '1' integrates to give x.

$$= \left[\frac{x^4}{4} - 2x^3 + 2x^2 + x\right]_0^1$$

Simplifying here will make the next steps easier.

$$= \left(\frac{1^4}{4} - 2(1)^3 + 2(1)^2 + 1\right) - 0 \quad (✓)$$

Substitute the upper and lower limits. Substituting a lower limit of zero will often give a bracket, which becomes zero. Make sure you include the '– 0' to demonstrate to the examiner that you have worked this out, and have not just ignored the lower limit.

$$= \frac{1}{4} - 2 + 2 + 1$$

$$= \frac{1}{4} + \frac{4}{4}$$

$$= \frac{5}{4} \text{ units}^2 \quad (✓)$$

Remember to give square units since it is an area you have found.

Hint

When the area you are trying to find lies entirely below the x-axis you should expect a negative answer when you integrate. There are different ways of dealing with this but whatever method you use you must communicate clearly to get all the marks.

(b)

$$\int_1^2 (x^3 - 6x^2 + 4x + 1)\,dx \quad (\checkmark)$$

Write out the **integral** with the new limits of 1 and 2. The value of this integral will be negative but will allow you to find the shaded area below the x-axis.

$$= \left[\frac{x^4}{4} - 2x^3 + 2x^2 + x\right]_1^2$$

You don't need to re-do the integral, just use your (simplified) result from part (a) and remember to change the limits.

$$= \left(\frac{2^4}{4} - 2(2)^3 + 2(2)^2 + 2\right) - \left(\frac{1^4}{4} - 2(1)^3 + 2(1)^2 + 1\right)$$

Substitute the upper and lower limits then subtract.

$$= (4 - 16 + 8 + 2) - \left(\frac{5}{4}\right)$$

Simplify. You don't have to work out the second bracket because you already figured it out in part (a).

$$= -2 - \frac{5}{4}$$

$$= -\frac{8}{4} - \frac{5}{4}$$

Take care with the fractions and negatives. Use as many steps here as you need.

$$= -\frac{13}{4} \quad (\checkmark)$$

We expected a negative answer here because the area lies entirely below the x-axis.

Second area $= \frac{13}{4}$ units^2

To get full marks you must make a clear statement about the area. Don't be tempted to write something like $-\frac{13}{4} = \frac{13}{4}$ or you will be penalised.

Total area $= \frac{13}{4} + \frac{5}{4}$ units^2

$= \frac{18}{4}$

$= \frac{9}{2}$ units^2 $(\checkmark)$

Add the two individual areas to obtain the total area.

Finding the area between a straight line and a curve

One of the main uses of integration is to find areas of shapes that would otherwise be impossible (or very difficult) to find. In many questions you will first be required to find the points where straight lines and curves **intersect**. The x-coordinates of these points will become the 'limits' of your integration. When you create your integral, don't forget to include the dx.

Example 19.4 

Q The line with equation $y=2x+3$ is a **tangent** to the curve with equation $y=x^3+3x^2+2x+3$ at A(0,3), as shown in the diagram.

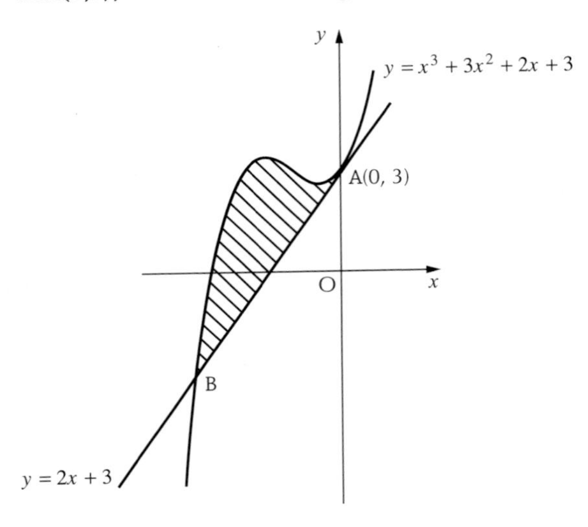

The line meets the curve again at B.

Shown that B is the point (–3, –3) and find the area enclosed by the line and the curve. **7**

A

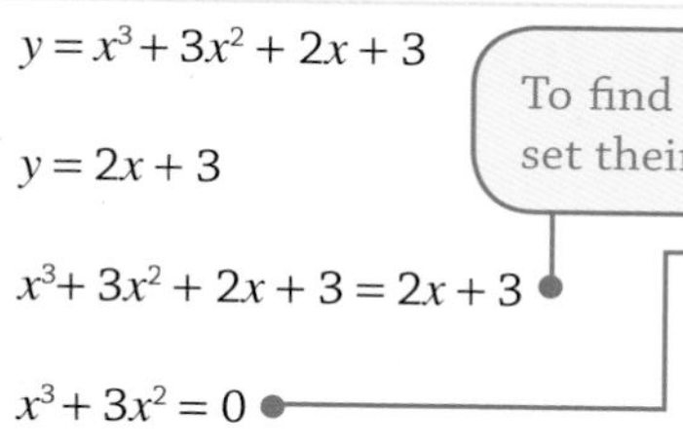

$y = x^3 + 3x^2 + 2x + 3$

$y = 2x + 3$

$x^3 + 3x^2 + 2x + 3 = 2x + 3$

To find where the line and the curve meet, set their equations as equal to each other.

$x^3 + 3x^2 = 0$

Simplify to the form ... = 0.

$x^2(x + 3) = 0$

Factorise. Here a common factor does the job, although in other questions you might have to use synthetic division. See Chapter 12 (Polynomials) for more on factorising and solving using synthetic division.

$x = 0$ and $x + 3$

$x = 0$ and $x = -3$ (✓)

Set both factors equal to zero and solve to find the x-coordinates of the points where the line and the curve meet.

When $x = -3$, $y = 2(-3) + 3 = -3$

Therefore the curves meet at B(–3, –3) (✓)

We already know from the question that the line and curve meet at A(0, 3). Substitute ($x = -3$) into the **linear** equation to find the corresponding y-value and show that B has coordinates (–3, –3).

$\int_{-3}^{0} (x^3 + 3x^2 + 2x + 3) - (2x + 3)\,dx$ (✓)(✓)

Integrate 'upper' – 'lower' between the limits of $x = -3$ and $x = 0$. Don't forget to include the dx.

$\int_{-3}^{0} (x^3 + 3x^2)\,dx$

Simplify.

$= \left[\frac{x^4}{4} + \frac{3x^3}{3}\right]_{-3}^{0}$

Integrate each term by adding one to the index then dividing by the new index. The dx goes at this stage.

$= \left[\frac{x^4}{4} + x^3\right]_{-3}^{0}$ (✓)

Simplify.

$= \left(\frac{0^4}{4} + 0^3\right) - \left(\frac{(-3)^4}{4} + (-3)^3\right)$ (✓)

Substitute the upper and lower limits then subtract. Use brackets for any negative values.

$= 0 - \left(\frac{81}{4} - 27\right)$

Simplify.

$= 0 - \left(\frac{81}{4} - \frac{108}{4}\right)$

Write 27 as $\frac{108}{4}$ to make simplification easier.

$= 0 - \left(-\frac{27}{4}\right)$

Simplify, take care with negatives.

$= \frac{27}{4}$ units^2 (✓)

Remember to give square units since it is an area you have found.

Determining the optimal solution for a given problem (optimisation)

These questions are easy to recognise once you have seen a few of them. You will very likely be asked to find the maximum or minimum value of some relationship and you must remember that differentiation is the method required. Optimisation questions are usually set in a real life physical context and will often require you to use your knowledge of shapes and common formulae for perimeter, area and volume. See Chapter 6 (Essential skills) for some important properties of shapes and see Chapter 7 (Formulae) for a list of useful formulae that are not provided in the exam.

Example 19.5

Q An open water tank, in the shape of a triangular prism, has a capacity of 108 litres. The tank is to be lined on the inside in order to make it watertight.

The triangular cross-section of the tank is right-angled and isosceles, with equal sides of length x cm. The tank has a length of l cm.

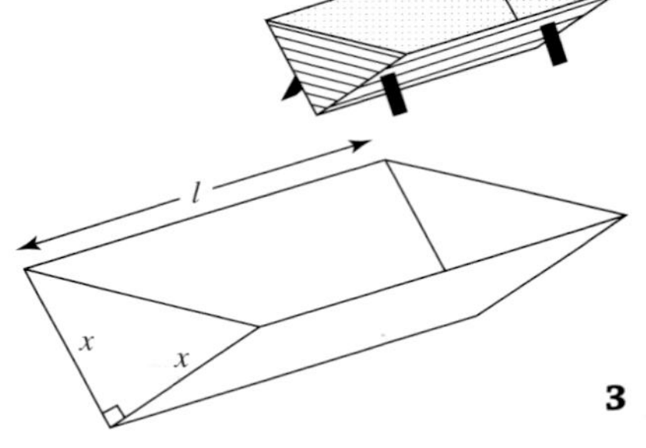

(a) Show that the surface area to be lined, A cm^2, is given by $A(x) = x^2 + \frac{432000}{x}$. **3**

(b) Find the value of x which minimises this surface area. **5**

A (a)

Surface Area $= 2 \times \frac{1}{2}x^2 + 2x \times l$ (✓)

The surface area of the tank comprises two triangular end sections plus two rectangular side pieces.

Hint

In many "show that" questions you don't have to be able to complete the "show that" part to be able to complete the rest of the question. This is because the result you need for the rest of the question is already given to you. In this case, all five marks from part (b) are available to you, even if you can't complete part (a).

Surface Area $= x^2 + 2xl$

This is an **expression** for surface area in terms of x and l. We want an expression that contains x only. We must now use the other information given in the question to create a substitution for l.

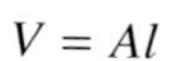

$V = Al$

Find an expression for volume. Remember that the volume of a prism = cross-sectional area × length.

$V = \frac{1}{2}x^2 l$

The cross-section is a right-angled triangle with area $= \frac{1}{2} \times \text{base} \times \text{height}$, i.e. $A = \frac{1}{2} \times x \times x = \frac{1}{2}x^2$.

$\frac{1}{2}x^2 l = 108000$

Equate your expression for volume with the value for volume given in the question. Change the volume from 108 litres to 108000 cm³ since the other dimensions are given in centimetres.

$l = \dfrac{108000}{\frac{1}{2}x^2}$ (✓)

$l = \dfrac{216000}{x^2}$

Rearrange and simplify to give an expression for length in terms of x.

$A(x) = x^2 + 2x\left(\dfrac{216000}{x^2}\right)$

Substitute the expression for length to create an expression for surface area exclusively in terms of x.

$A(x) = x^2 + \dfrac{432000}{x}$ (✓)

Simplify to give the exact expression asked for in part (a).

(b)

$A(x) = x^2 + \dfrac{432000}{x}$

$A(x) = x^2 + 432000x^{-1}$ (✓)

Prepare the function for differentiation by writing each term in the form ax^n.

$A'(x) = 2x - 432000x^{-2}$ (✓)

Differentiate each term by "multiplying by the index then subtracting one from the index".

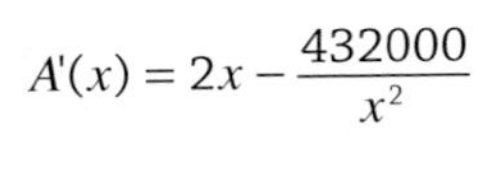

$$A'(x) = 2x - \frac{432000}{x^2}$$

Writing the derivative with positive indices will often make the next steps easier.

$$2x - \frac{432000}{x^2} = 0 \quad (✓)$$

Stationary points occur when the derivative is equal to zero.

$$2x^3 - 432000 = 0$$

Begin to solve for x by multiplying both sides by x^2.

$$2x^3 = 432000$$

$$x^3 = 216000$$

$$x = 60 \quad (✓)$$

Solve to give the x-coordinate of the stationary point of the function.

x	(59) →	60	(61) →
$A'(x)$	–	0	+
Shape	╲	—	╱

(✓)

Use a **nature table** to investigate the gradient of the curve on either side of the stationary point. Knowing if it is positive or negative will tell you the shape of the curve.

Substitute sensible values into your derivative and indicate in the table whether the result is positive or negative. You already know the gradient will be zero at the stationary point.

The surface area of the tank is at its minimum when $x = 60$ centimetres.

Hint

Make sure you include all the information shown above in your own nature tables, otherwise you risk losing marks. If your function involves a variable (letter) other than x, you must make sure you use that variable in your nature table.

Glossary of mathematical terms and symbols

e	An irrational number that appears in many areas of Mathematics. It is approximately equal to 2.718... and can be found on most scientific calculators.
θ	The Greek letter, Theta, often used to denote an unknown angle.
α	The Greek letter, Alpha, used to denote an unknown angle.
Altitude	An altitude is a line drawn from one vertex of a triangle to meet the opposite side at 90°. All triangles have three altitudes. The altitude is also known as the height of a triangle.
Amplitude (Trig)	The amplitude of a trigonometric function is half the distance between the maximum and the minimum values, i.e. $\frac{\text{max} - \text{min}}{2}$.
Base	The base represents the number of digits in a number system, e.g. base 10 has 10 digits (0 – 9) while base 2 has two digits (0 and 1). Bases are used in exponentials and logarithms to indicate which number system is being used, e.g. in $y = 4^x$ and $y = \log_4 x$ the base is 4.
Basis vectors i, j and k	Three mutually perpendicular vectors each with a magnitude of one. The basis vectors are used as 'building blocks' to represent other vectors.
Bisect	To cut or divide into two equal parts.
Centre	The middle of an object. A circle has one centre, however, a triangle has many different centres all with different properties.
Chain rule (Differentiation)	Used to differentiate composite functions, e.g. the derivative of $f(g(x))$ is given by $f'(x) \times g'(x)$, i.e. differentiate the 'outer' function then *multiply* by the derivative of the 'inner' function.
Chain rule (Integration)	Used to integrate composite functions, e.g. the integral of $f(g(x))$ is found by integrating the 'outer' function, $f(x)$, then *dividing* by the derivative of the 'inner' function, $g(x)$.
Chord	A line connecting two points on the circumference of a circle. When a chord passes through the centre of the circle it is called the diameter.
Coefficient	A number multiplied with a variable in an algebraic term, e.g. 5 is the coefficient of x in the expression $5x - 7$. A coefficient of one (or negative one) would not normally be written, e.g. -1 is the coefficient of x in the expression $7 - x$.

Collinear Three or more points that lie on the same straight line are said to be collinear. Two points will always be collinear.

Component One 'part' of a vector. A vector in 2 dimensions will have 2 components, usually x and y. A vector in 3 dimensions will have 3 components; x, y and z. Vectors are written in component form to make them easier to work with, e.g. $\begin{pmatrix} 4 \\ 1 \\ 7 \end{pmatrix}$ is a 3-dimensional vector in component form.

Composite function A composite function is formed when two (or more) functions are applied to a variable one after the other. Applying the function $f(x)$ to the result of the function $g(x)$ creates the composite function $f(g(x))$.

Concurrent Three or more lines are concurrent if they pass through the same point.

Congruent Congruent shapes are exactly equal in shape and size.

Constant A number (or letter that represents a number) on its own in an algebraic expression, e.g. 8 is a constant in the expression $2x + 8$.

Constant of integration A numerical value which is added to all indefinite integrals, initially represented by '$+ C$'. In some cases it is possible to evaluate C if additional information is given.

Definite integral An integral with limits. In this case the integration will simplify to a numerical value, e.g. $\int_1^5 (x^2 + 4x)\,dx$.

Derivative A measure of the gradient (or rate of change) of a function for a given value of x.

Derived function The result obtained by differentiation, commonly written as $f'(x)$ or $\frac{dy}{dx}$.

Differentiation The name given to the process used to find a derivative. For expressions in the form ax^n, the derivative is nax^{n-1}.

Directed line segment A line of fixed length that connects two points in a specific direction, e.g. $\overrightarrow{AB}$.

Discriminant (nature of roots) $b^2 - 4ac$ used to determine the nature of the roots of a quadratic function.

Domain (of a function) The set of input values for which a function is defined, i.e. the numbers that can go 'in' to a function and give a real number output. Dividing by zero or taking the square root or logarithm of a negative should be avoided.

Exact value An exact value answer should not be rounded. Simplified fractions and surds are exact values, e.g. $\sqrt{2}$, $\frac{1}{3}$ and $\frac{\pi}{6}$ are all exact values.

Exponent (index) A small number written above and right of a number to indicate how many of the base number are to be multiplied together, e.g. 4 is the exponent in 2^4 and tells us that four 2s are to be multiplied, i.e. $2^4 = 2 \times 2 \times 2 \times 2 = 16$.

Exponential (function) A function containing one or more quantities raised to an exponent, e.g. $f(x) = 2^x$. Unless they have been transformed in some way, graphs of exponential functions always pass through the point (0,1).

Expression A collection of variables (letters) and constants (numbers) connected by mathematical operations. Expressions do not usually contain an equals sign, e.g. $\cos(x - 30)°$ and $3x^4 + 5x - 1$.

Factor Numbers or algebraic expressions which, when multiplied together, create a given product, are called factors of that product, e.g. 4 and 3 are two factors of 12, while $4x$ is one factor of $12x^2$.

Factorise Factorising an expression involves rewriting it as a product of its factors, e.g. $14x^2 + 21x$ can be written as $7x(2x + 3)$by removing a common factor of $7x$ from both terms in the original expression.

Frequency (trig) The number of cycles a trig function completes in a given interval, usually 360 degrees or 2π radians for the sine and cosine functions.

Function A mathematical 'rule' connecting input (usually x) to output (usually y). The notation $f(x)$ is often used when defining a function, e.g. $f(x) = 3x^3 + 5$ and $g(x) = 4 \sin x - 2$ are functions.

Gradient (derivative) The steepness of a curve at a given point can be found by evaluating the derivative of the function using the x value of the given point.

Gradient (of a line) A measure of the steepness of a line relative to the x-axis. Gradient has a numerical value and is found by dividing vertical change by horizontal change.

Image (of a point) When a graph of a function has undergone a translation, i.e. it has been transformed in some way, a point, P, on the original function will have an image, P′, on the transformed graph.

Indefinite integral An integral without limits. Here the result will be an expression and will include the constant of integration, $+ C$.

Index See Exponent.

Integral An infinite sum of rectangles of infinitely small width.

Integration Used in Calculus to find areas and volumes of shapes when standard geometrical rules and formulae are not sufficient. Integration is the reverse process of differentiation.

Intersection (with axes or between curves/lines) A single point where two or more lines cross or meet.

Inverse (function) A function which 'reverses' or 'undoes' another function. For a function $f(x)$, the inverse function would be written as $f^{-1}(x)$. If a value is used as input to a function and the result is then input to the inverse function (or vice-versa), the final output will be the original value, i.e. $f^{-1}(f(x)) = f(f^{-1}(x)) = x$.

Irrational (number) A real number that can't be made by dividing two integers, e.g. π and e are irrational numbers, so are surds like $\sqrt{2}$.

Irrational (roots) The roots of a quadratic function will be irrational if the discriminant is any number other than a square number.

Limit (of a sequence) If it exists, the limit of a sequence is a value that the terms in a sequence 'tend to', i.e. as the sequence continues the terms approach the same value. For sequences of the form $u_{n+1} = au_n + b$, a limit will exist if $-1 < a < 1$. The limit of a sequence can be found using the formula $L = \frac{b}{1-a}$.

Limits (of integration) When evaluating a definite integral, the limits of integration will be the lowest and highest values of x for which the integral exists, e.g. in $\int_2^5 3x^2\,dx$, the upper limit is 5, and the lower limit is 2.

Linear (function) A function where the highest power of the variable is one. E.g. $f(x) = 3x + 5$ is a linear function which, when graphed, will produce a straight line.

Logarithm The power to which a number must be raised to give a certain value. Logarithms can be used to solve exponential equations, e.g. the solution to $3^x = 9$ is $x = 2$, using the fact that 3 must be raised to the second power to give 9. This relationship can also be written as $\log_3 9 = 2$.

Logarithmic function A function of the form $y = \log_b x$ is a logarithmic function. Here b represents the base of the function. Base 10 is used in our decimal number system, although base e is also common. The logarithmic function is the inverse of the exponential function, i.e. logarithms undo exponentials and vice-versa.

Magnitude (of a vector) The size or length of a vector. For the three dimensional vector $\mathbf{a} = \begin{pmatrix} x \\ y \\ z \end{pmatrix}$, the magnitude, $|\mathbf{a}|$, is found using $|\mathbf{a}| = \sqrt{x^2 + y^2 + z^2}$.

Median (of a triangle) A line from one vertex of a triangle to the midpoint of the opposite side. Triangles have three medians which lie inside the triangle. The medians are **concurrent** and meet at the centroid of the triangle.

Natural logarithm A commonly occurring logarithm in science and engineering where the base is the irrational number, e. A natural logarithm will often be written as $\ln x$ or $\log_e x$.

Nature table Used to determine the nature of the stationary points of a function, i.e. maximum, minimum or point of inflection. The nature is found by examining the gradient of the function on either side of a stationary point. An alternative to using a nature table is the second derivative test.

Optimisation The process of finding a maximum/minimum value of a function, usually requiring differentiation.

Parabola A $\cup$ or $\cap$ shaped curve produced by drawing the graph of a quadratic. If the coefficient of x^2 is < 0, the graph will be $\cap$ shaped with a maximum turning point. If the coefficient of x^2 is > 0, the graph will be $\cup$ shaped with a minimum turning point.

Parallel Lines which are the same distance apart along their entire length. Parallel straight lines have equal gradient.

Period For a repeating function like $y = \sin x$ or $y = \cos x$, the period is the horizontal length of one complete wave (or cycle).

Perpendicular lines Lines meeting at 90°. The gradients of perpendicular lines multiply to give -1.

Perpendicular vectors Vectors where the angle between them is 90°. If two vectors are perpendicular their scalar product will equal 0, i.e. $\mathbf{a.b} = 0$.

Phase angle For functions like $f(x) = \sin(x + d)$, the constant, d, controls how far the graph of $y = \sin x$ 'shifts' horizontally along the x – axis. If $d > 0$, the shift will be to the left. If $d < 0$, the shift is to the right.

Point of inflection (rising/falling) A stationary point on a function which is neither a maximum nor a minimum. The gradient of a curve goes from positive to zero and back to positive at a rising point of inflection, and from negative to zero to negative at a falling point of inflection. Points of inflection can be investigated using a nature table.

Position vector The components of the 'journey' from the Origin to a given point, e.g. the position vector of the point (3,4) is $\begin{pmatrix} 3 \\ 4 \end{pmatrix}$.

Radian An alternative angle measure. One radian is the angle swept out by the radius of a circle as it travels along an arc with the same length as the radius. π radians = 180 degrees, making one radian equal to a little less than 60 degrees.

Range (of a function) The 'output' values of a function. This can be thought of as all the y – values between the maximum and minimum points of the function. For many functions, the range is all the real numbers.

Rate of change The gradient of a line is a rate of change; it is a numerical value describing how quickly the output of a function changes with the input. In other words, the change in y divided by the change in x or $\frac{dy}{dx}$. Rate of change is the derivative. One example is velocity, which is the rate of change of displacement with time.

Ratio Usually given as a fraction, a ratio is a way of comparing two numbers using division.

Rational (number) A number that can be made by dividing two integers. Recurring decimals like 0.333... and 0.666... are rational since 0.333...can be written as $\frac{1}{3}$ and 0.666...can be written as $\frac{2}{3}$.

Rational (roots) The roots of a quadratic function will be rational if the discriminant is a square number, e.g. 1, 4, 9 etc...

Recurrence relation An equation which 'builds' a sequence by creating the next term from the current term. For a recurrence relation to be properly defined, a starting (initial) term should be given.

Roots (of a function) The values where the graph of a function crosses (or touches) the x-axis. Also thought of as the x value(s) for which $f(x) = 0$. The roots are also known as zeros, or solutions.

Scalar Often used in the context of vectors, a scalar is a real number.

Scalar product (vectors) The result of multiplying the magnitude of two vectors by the cosine of the angle between them. The scalar product is a real number.

Second derivative (test) The result of differentiating a function twice, e.g. if $f(x) = x^3$ then $f'(x) = 3x^2$ and $f''(x) = 6x$. The second derivative can be used instead of a nature table to determine whether a stationary point is a maximum or a minimum.

Sequence (of numbers) An ordered set of numbers (elements) generated using a fixed rule, e.g. the rule $T_n = 3n + 1$ generates the sequence 4, 7, 10, 13,... where T_n is the term at position n in the sequence.

Set (of numbers) A collection or group of numbers that have a common property, e.g. {2, 3, 5, 7...} is the set of prime numbers. The set of real numbers, denoted by $\mathbb{R}$, contains all the rational and irrational numbers.

Stationary point(s) The coordinates of the point(s) on the graph of a function where the gradient equals zero, i.e. $\frac{dy}{dx} = 0$.

Strictly increasing/ decreasing A function where the gradient, $\frac{dy}{dx}$, is always positive is said to be strictly increasing. If the gradient is always negative the function is strictly decreasing. Note that a function is neither increasing nor decreasing at any point where its gradient is zero.

Tangent (to a curve)	A line that touches a curve without cutting through it. At the point of contact the gradient of the tangent line is equal to the gradient of the curve.
Turning point – maximum/minimum (of a function)	A point on the graph of a function where the gradient of the graph becomes zero as it changes from positive to negative (maximum turning point) or from negative to positive (minimum turning point).
Unit vector	A vector with magnitude equal to one.
Values – maximum/ minimum (of a function)	The highest (max) or lowest (min) y – coordinate on the graph of a function. Max/min values are single numbers, not coordinates.
Vector	An object that can be defined by two quantities; magnitude and direction. Displacement, velocity and force are all examples of vectors.

Additional information

Symbols, terms and sets

The content listed below is not examinable. However, you are expected to understand:

The symbols: $\in$, $\notin$, $\{\}$
The terms: set, subset, empty set, member, element
The conventions for naming sets, namely:

- $\mathbb{N}$, the set of natural number, i.e. {1,2,3, . . . }
- $\mathbb{Z}$, the set of integers, i.e. { . . .-3,-2,-1,0,1,2,3, . . . }
- $\mathbb{Q}$, the set of rational numbers
- $\mathbb{R}$, the set of real numbers
- Notes: the set of whole numbers are a subset of the integers and include all the natural numbers, plus zero, i.e. {0,1,2,3, . . . }